Claim the Moon!

A Blueprint to Checkmate China, Recover Every Cent We Have Spent on Space Exploration, and Guarantee Our Future Prosperity

Claim the Moon!

A Blueprint to Checkmate China, Recover Every Cent We Have Spent on Space Exploration, and Guarantee Our Future Prosperity

WALT CROSS

DIRE WOLF BOOKS
STILLWATER, OKLAHOMA

A Division of
CROSS PUBLICATIONS
502 E. Liberty Avenue
Stillwater, OK 74075-2630

First Edition 2020

Copyright © 2020 by Walt Cross. All rights reserved. No part of this work may be reproduced or transmitted, in any form or by any means, electronic or mechanical, including photocopying, recording, or any information storage and retrieval system without permission in writing from the publisher and author.

Cover created by Walt Cross, copyright © 2020.
Cataloging Data

Cross, Walter L.
Claim the Moon; A Blueprint to Checkmate China and Recover Every Cent We Have Spent on Space Exploration

1. The Moon 2. Outer Space 3. Space Exploration. 4. Space Corp/Space Force. 5. Colonizing outer space. 6. Planetary Science. I. Title

MANUFACTURED IN THE UNITED STATES OF AMERICA

DEDICATION

This work is dedicated to the memory of John Cornelius Houboult, NASA aerospace engineer.
April 10, 1919 to April 15, 2014

Thank you John, without you we would not have achieved the Moon. If you don't know the story of John Houboult, you might like to take the time to find out.

Photograph courtesy NASA.

Space is indeed the final and infinite frontier, always expanding.

Walt Cross

'And I think it's gonna be a long long time 'Till touchdown brings me back around to find…I'm a rocket man…'

-Elton John, 1972

Table of Contents

Foreword		1
Chapter 1	**The Moon**	5
	Formation of the Moon	
	Lunar Terrain	
	Exploration	
	Successful Moon Missions 59-69	
	Successful 1970s Moon Missions	
	Successful 21st Century Moon Missions	
	Interior of the Moon	
	Moon Mascons	
	Moon Magnetic Umbrellas	
	The Moon, Stepping Stone to Mars	
	Layered Regolith in the Von Karman Crater	
Chapter 2	**We Return to the Moon**	29
	Our Biggest Past Moon Error	
	NASA's Artemis Program	
	Space Policy Directive 1	
	White House Moon Spending Plan 2021	
	NASA Statement on FY 2021 Funding	
	Orion Spacecraft	
	Funding Artemis	
	Moon Nuclear Power	
	Kilopower Project	
	Nuclear Propulsion Spacecraft	
	Nuclear Cislunar Rocket	
	Cosmic Radiation	
	New Spacesuit	
	Lunar Landers	
	Walking and other Moon Rovers	

Project Farside

Chapter 3 **First Commercial Astronauts** 60
Astronaut's Backgrounds
Women in Space and on the Moon
First Generation of Artemis Astronauts
Christa Koch Sets Space Flight Record
Space Flight Test of the Starliner
Blue Origin's New Shepard
SpaceX Crew Dragon Test Success
He Likes Rockets
Possible Manned Mission in May 2020
NASA Seeks Students for M2M XHAB

Chapter 4 **Space Corps/Space Force** 87
Space Command
45th Space Wing
Military Astronauts
KBR Astronaut Training
Congress Questions USSF Commander
Space Development Agency
Space Force – Space Academy
China Reaction to U.S. Space Force
The Cost of a USSF Moon Base
Space Weapons

Chapter 5 **Regulating Outer Space** 118
Death in Space and on the Moon
Moon Invasion and Contamination
First Orbit Cleanup Spaceship
Japan enters Orbit Cleanup Effort
ClearSpace-1 Mission
A Space Scare over Pittsburgh

End of Manned Spaceflight Drought
Emergency Earth Rescue

Chapter 6 **Return on Our Space Investment**   129
The New Gold Rush
Moon Facilities Available for Int'l Use
Lockheed Martin Space Growth
Lunar Natural Resources
Trillion Dollar Metal World of Psyche
Atomic Probe of Moon Dust
Moon Penal Colony
First Satellite to Satellite Refueling

Chapter 7 **Getting Off to a Fast Start**   145
Renewal of NASA Manned Spaceflight
New U.S. – Russia Space Race
Upcoming Space Missions for 2020
SpaceX Requests Early Starship Launch
SpaceX to Speed Starship Production with a Cautionary Note
Space Tourism 2020
A Clock Work Orange on Venus

Chapter 8 **Claiming the Moon**   159
Land Claims in the U.S.
Mining Claims in the U.S.
Staking a Claim
How to Stake a Moon Claim
Outer Space International Treaties
Ipso Facto
The Space Act of 2015

Chapter 9	**Justifying Our Claim** Prince Rupert's Land Columbus' Claim	174
Chapter 10	**Why We Must Lay Claim Now** Mission Clementine Bistatic Radar Experiment NASA's LCROSS Mission Possible Claim Challengers	183
Chapter 11	**Threats to U.S. Space Programs** Air Force Issues China, Russia Warning Communist Chinese Hacking and Espionage Chinese Hackers Take Control of ISS Scientist Hides China Links from NASA Russians Threaten U.S. Satellite Prediction; a China Surprise Moon Landing	202
Chapter 12	**The Moon, Gateway to the Solar System and Beyond**	211
Chapter 13	**The World's Reaction** You Will Remember Me for Centuries	215
Epilogue	 National Claim Movement Letter Remembering the Challenger Crew	219
About the Author		223

Foreword

"We choose to go to the Moon...We choose to go to the Moon in this decade and do the other things, not because they are easy, but because they are hard; because that goal will serve to organize and measure the best of our energies and skills, because that challenge is one that we are willing to accept, one we are unwilling to postpone, and one we intend to win, and the others, too."

When President John F. Kennedy gave this emotional speech at Rice University in Houston on September 12, 1962, do you think he intended it as an empty gesture? No, of course you don't. Like all the rest of us that heard that original speech or have heard it since, you know he meant every word. And so did the original seven astronauts of that time along with the entire nation. And, we overcame all the challenges and stepped on the Moon less than seven years later! But by 1972 we withdrew our last manned Moon mission and made all of what came before a virtual empty gesture. Why do I say that? Because we failed to claim the Moon, we failed to make it ours when we had the chance. It's not too late to change that, and this book is all about changing that.

I and a good number of other Americans view the time after 1972 until the most recent of times as a glorious time of robotic unmanned space flight, but a waste of time during that same era for manned space flight. I understand that we gained a great deal of knowledge from the space shuttle flights and the occupation of the International Space Station. But in many ways it felt as though we were standing still.

That part of the speech that says the challenge will measure our energies and skills was certainly correct. And now we are in a similar situation, where we need all our energies and skills, added to our much greater knowledge, to do the same feat again but do it in a manner that lets us stay forever this time. We now must put a human presence on a second celestial body of our solar system, and stay there!

Ever since the beginning of the United States, the idea of a seemingly never ending frontier open to unlimited expansion has been an element of what it means to be American. The edge of space, with its endless call for exploration and its infinite halls of room for expansion is now become that frontier.

We are but a few short years from the beginning of the end of our story as simply the People of Earth. This milestone is set for the year of 2024 when we return to the Moon to stay. From that day and forever more when we write about it, the word Moon, meaning the celestial body that circles our Earth, must be capitalized. We should do no less for the Second Home of Mankind.

It is a milestone that will mark the start of the greatest human diaspora since the move en masse from Europe to the American continents. This will not be a fast change because we must go forward carefully into a hostile and infinitely dangerous environment, the environment of free and open space.

The circle of life will grow much larger to encompass the Moon, the asteroids, and eventually the planet of Mars and then, even more. The struggle to survive and thrive in the realm of space is a challenge to be met with a bold directness by the United States and its space faring allies.

The first goal must be to gain an entire world we call *the* Moon. Yes, gaining the Moon equates to nothing less than adding an entire continent to the United States of America! And we must move directly to do this without delay. We need a Moon Express to go there and never leave again. We must take advantage of our technological edge in the exploration of space and the Moon. We are foremost among the space faring nations and we must remain foremost and increase our presence exponentially on the Moon.

To do this we must lay our claim to the Moon, the entire Moon, and we must do it now. We must be prepared to defend that claim by force if necessary, and there is little doubt that it will be necessary. Even now we are creating the force to do that, the United States Space Force.

This must be done now for the future of humanity. We must not let the Moon and its vast resources fall under the control of some oppressive or dictatorial political power such as Communist China. What follows is my vision of how we do that.

The original seven Mercury astronauts. Courtesy of NASA.

CHAPTER 1
THE MOON

The Moon is the protector of Earth. It has had the unenviable honor for literally billions of years of keeping the fury of the cosmos at bay. When the Earth and the Moon first formed and cooled our planetary sized satellite provided stability. It kept Earth's axis wobble small and saved it from rolling over on its side. This was a first necessary step in the birth of life on Earth and resulted in the seasons as we know them today.

This stability was and is absolutely necessary for a viable climate to take hold and nurture life on the surface of our planet. It also created the tides that are the breath of life for the Earth. At a distance of 239,000 miles the Moon is the closest celestial body to Earth and to humans. It is so close that we can reach it in a mere three days of travel by spacecraft.

This oh-so-convenient location is an open invitation for us to visit it, explore it, adapt to it, and learn to live upon its surface or perhaps just below its surface.

We, the people of the United States have done all of these things, and we need to do them again on a bigger scale both physically speaking and spiritually.

In the beginning of the Earth and Moon relationship, Mother Earth was a harsh parent indeed to the young Moon. She scorched her only child with a fiery breath for millennia and jerked her to a violent standstill forcing the Moon to show only one side of her face, that stark and scarred yet beautiful visage humans so revere.

She is also our shield maiden, providing cover for us from many of the asteroids that pummeled both the Earth and Moon during the Late Heavy Bombardment[1], protecting Earth from much of the cosmic violence.

From our Earthly home we always see the same side of the Moon because the Moon spins on its axis, with a speed that places it in a synchronous rotation with Earth. The Moon is said to be tidally locked to the Earth. That is why we can never see the far side of the Moon unless we go there. It wasn't until October 7, 1959 that we saw a grainy picture of the far side of our Moon taken by the Soviet Union's Luna 3 probe.

Image of the late heavy bombardment.

---

[1] From about 4.5 to 3.8 billion years ago, failed planets and smaller asteroids slammed into larger worlds, damaging and scarring their surfaces.

Formation of the Moon

Over the centuries there have been a number of theories about how the Moon was formed. Here are some of them.

There is the Theia Impact theory that suggests that the Moon formed out of the debris left over from a collision between Earth and an astronomical body the size of Mars billions of years in the past. When I first read this theory I recalled a game we used to play called balderdash. Balderdash is a reference to senseless talk or writing. And that's how I felt about this theory. It's just not provable. First it suggests a fictional planet that someone made up and named Theia existed. Then it suggests this fictional planet crashed into the Earth…I think you see my point, it's ludicrous on the face of it and is too complicated. However, you may be interested enough to look this one up.

Recently after 'new research' the Theia theory was changed just a bit by referring to different percentages of elements found on Earth and the Moon that to the researcher's mind increased the facts pointing to a Theia impact. However, I still don't buy it.

Another theory just as implausible hypothesizes that the Earth had not one, but two moons in the beginning. It's yet another case of complicated balderdash. But, I do applaud their imagination.

No, let's keep it simple shall we? That's the way these things usually are in nature. I have always been a proponent of the suggestion that the Earth and the Moon formed together at the same time and near one another. And, there is proof of this.

In 2011 two astronomers Dr. Jason Wright and Dr. Arpita Roy looked into why the crust of the far side of the Moon is thirty miles thicker than the near side.

Wright was studying exoplanets when he found one that was so close to its sun, that it was being cooked! The star was acting like a blow torch, keeping the planet in a permanent state of molten lava. The planet was tidally locked to the star, always presenting only one side of itself to the conflagration of solar energy.

He surmised that the same kind of thing could have happened to the Moon. When they formed, both the Moon and the Earth were in a molten form. The Moon was a mere 15,000 to 20,000 miles away (as compared to its present location of 239,000 miles away).

This means the Moon was close enough to suffer the same cooked planet syndrome that Jason Wright observed. The Earth, acting like a blow torch, pushed molten Moon material to the far side, causing the 30 mile thicker crust to form and cool on the far side where Earth's fiery breath could not reach.

Consulting his colleague Arpita Roy, Jason asked her to use their powerful computer simulation program to find out how long it took for the moon to become tidally locked to the still flaming Earth.

Dr. Arpita was surprised to learn that it only took a hundred days for the spinning Moon to slow down and then stop, tidally locked from that distant time of 4.5 billion years to today.

This strongly suggests that the molten Earth and Moon were not only very close together, but the Moon displayed one side of its face to its host planet from the very start.

So, I think this 'Cooked Moon'[2] theory checks all the boxes on how the Moon was formed. At least I think it's the truth, for now.

Lunar Terrain

The dark areas are called maria (MAH-ria), Latin for seas that are actually impact basins full of ancient lava beds formed billions of years ago. These areas may have formed during the aforementioned heavy bombardment. The light areas

---

[2] This is my phrase. I don't think this theory has a name yet. Oh, I guess I just gave it one.

of the Moon's surface are known as the highlands as they were too high for the lava flows to climb them.

Overall the surface boasts rocks of varied composition and age. An impact history of the Moon is revealed by the many craters that have existed for as long as billions of years to just yesterday. The craters suffer little erosion over time.

This excellent and clear photograph of the Moon's far side was taken in April, 1972 by Apollo 16 as it made its journey around the Moon. (NASA)

What might be termed the Moon's surface soil is called regolith, a gray charcoal colored and powdered dust. Under this 'soil' is a hardened crust of broken and crushed rock termed the megaregolith.

Exploration

In 1959 the Soviet Union launched Luna 1, the first spacecraft to come close to the Moon. But sadly for the Russians, it totally missed the Moon.

They soon followed up with Luna II that same year. This time the spacecraft took a direct route to the Moon, sent back telemetry, and then impacted the lunar surface.

Since these attempts, as of April of 2019 seven nations have launched Moon missions. The U.S. sent several robotic missions from 1961 to 1967 to pave the way for our manned missions. These included both impact probes and lunar orbiters tasked to map the surface and find appropriate landing sites. The final missions were soft landers that, of course, soft landed on the Moon.

The first manned landing on the Moon; Apollo 11, was led by astronaut and mission commander Neil Armstrong, on July 20, 1969.

At the time I was a young soldier on my way back to Vietnam after spending R&R[3] with my wife in Hawaii. I landed back in Vietnam while Neil and Buzz Aldrin were landing on the Moon. I got to

---
[3] Rest and Recuperation.

watch the event on TV before returning to my unit in the field. Little did I know I would shortly meet and shake hands with Neil Armstrong.

In December of 1969 the Bob Hope Christmas Show came to Vietnam and I got to see it. Neil Armstrong[4] came along with the show and as mentioned, I was fortunate enough to meet him and shake his hand. As did, I might add, hundreds of other soldiers. But that didn't lessen my thrill at meeting him. I had followed NASA's space program as a young man with fascination from its very inception.

Our astronauts walked on the Moon and performed many tasks including driving a Lunar Rover, collecting Moon samples, and conducting scientific experiments. They eventually, during the ensuing Apollo missions, brought back more than eight hundred pounds of lunar soil and rocks for study on Earth.

Unfortunately, in a short sighted decision by the government, lunar exploration waned and did not resume until the 1990s. I say government, because NASA was ready to push on with an advanced Moon program. Other missions continued and those are looked at in detail later in this narrative.

---

[4] Neil Armstrong, the first man to step upon the Moon, died in August 2012 forty-three years after his famous mission.

Geologist-Astronaut Harrison Schmitt works next to a huge, split boulder at geology Station 6 on the sloping base of the North Massif during the Apollo 17 mission. Photograph courtesy of NASA.

However, in March of 2019 NASA Administrator Jim Bridenstine announced that U.S. astronauts will return to the surface of the Moon by 2024. This is an event that will be welcomed by millions of people in both of the United States and all around the world.

Also in March of 2019, The European Space Agency as well as China, Japan and India sent missions to the Moon. China landed two rovers on the surface, including the first-ever landing on the Moon's far side. In another first, Israel sent a spacecraft to land on the Moon in April. It managed to orbit the Moon but unfortunately was destroyed in an attempted soft landing.

Successful Moon Missions
1959 to 1969

Luna 1 USSR Flyby 1959	Partial Success[5]
Pioneer 4 USA Flyby 1959	Partial Success
Luna 2 USSR Impact 1959	Success[6]
Luna 3 USSR Flyby 1959	Success[7]
Ranger 7 USA Impact 1964	Success[8]
Ranger 8 USA Impact 1965	Success
Ranger 9 USA Impact 1965	Success
Luna 9 USSR Lander 1966	Success[9]
Luna 10 USSR Orbiter 1966	Success[10]
Surveyor 1 USA Lander 1966	Success[11]
Lunar Orbiter 1 USA Orbiter 1966	Success
Luna 12 USSR Orbiter 1966	Success
Lunar Orbiter 2 USA Orbiter 1966	Success
Luna 13 USSR Lander 1966	Success
Lunar Orbiter 3 USA Orbiter 1966	Partial Success[12]
Surveyor 3 USA Lander 1967	Success[13]
Lunar Orbiter 4 USA 1967	Partial Success[14]
Lunar Orbiter 5 USA 1967	Success

---

[5] First lunar flyby.
[6] First Moon impact.
[7] First pictures of lunar farside.
[8] First close-up pictures of the Moon.
[9] First soft landing on the Moon and first pictures from lunar surface.
[10] First lunar orbiter.
[11] First US soft landing and first US pictures from Lunar surface.
[12] Camera failure
[13] Visited by Apollo 12 crew on the Moon.
[14] Camera failure.

Surveyor 5 USA Lander 1967 Success
Surveyor 6 USA Lander 1967 Success
Surveyor 7 USA Lander 1968 Success
Luna 14 USSR Orbiter 1968 Success
Zond 5 USSR Flyby 1968 Success[15]
Zond 6 USSR Flyby 1968 Partial Success[16]
Apollo 8 USA Orbiter 1968 Success[17]
Apollo 10 USA Orbiter 1968 Success[18]
Luna 15 USSR 1969 Orbiter/Sample Return Partial Success[19]
Apollo 11 USA Orbiter/Lander 1969 Success[20]
Zond 5 USSR Flyby 1968 Success
Apollo 12 USA Orbiter/Lander 1969 Success[21]

---

[15] First live creatures to fly by the Moon.
[16] Animal cargo lost.
[17] First humans to orbit the Moon.
[18] Moon lander successfully tested in lunar orbit.
[19] Sample return attempt crashed on the Moon.
[20] First humans to land on the Moon and return safely to Earth.
[21] Second successful landing of humans on the Moon and return safely to Earth.

NASA's Dr. Wernher von Braun, John Casani and Dr. James Van Allen.

Apollo 12 commander Charles Conrad Jr. examines the robotic Surveyor.

Successful 1970s Moon Missions

Apollo 13 USA Orbiter/Lander 1970 Partial Success[22]
Luna 16 USSR 1970 Lander/Sample Success[23]
Zond 8 USSR Flyby 1970   Success
Luna 17 USSR Lander 1970   Success[24]
Apollo 14 USA Orbiter/Lander 1971 Success[25]
Apollo 15 USA Orbiter/Lander 1971 Success[26]
Luna 19 USSR Orbiter 1971   Success
Luna 20 USSR Orbiter/Sample 1972 Success
Apollo 16 USA Orbiter/Lander 1972 Success[27]
Apollo 17 USA Orbiter/Lander 1972 Success[28]
Luna 21 USSR Lander 1973   Success[29]
Mariner 10 USA Flyby 1973   Success[30]
Luna 22 USSR Orbiter 1974   Success
Luna 23 USSR Lander 1974   Partial Success
Luna 24 USSR Lander 1976   Success

---

[22] Crew in imminent danger of death due to failure of spacecraft. Moon landing mission canceled. Dramatic and surprising repairs were made in orbit and the crew returned safely to Earth.
[23] First robot to return sample to Earth.
[24] Delivered Lunokhod 1 rover to the surface of the Moon.
[25] Third successful landing of humans on the Moon and return safely to Earth.
[26] Fourth successful landing of humans on the Moon and return safely to Earth. Deployed PFS-1 orbiter mission launched in orbit.
[27] Fifth successful landing of humans on the Moon and return safely to Earth. Deployed PFS-2 orbiter mission launched in orbit, orbit decayed.
[28] Sixth successful landing of humans on the Moon and return safely to Earth.
[29] Delivered Lunokhod 2 rover to the surface of the Moon. Longest lived Moon rover.
[30] Studied the Moon on its way to Mercury.

Apollo 17 astronaut Harrison Schmitt, Apollo 17 collects lunar samples at the Taurus-Littrow landing site. Photograph and caption courtesy NASA.

In the 1980s it was a lonely Moon. No one visited.
A Record of *Successful* Moon Missions of the 1990s

Hiten Japan 1990 Orbiter/Impact   Success[31]
Clementine USA 1991 Orbiter   Success
Lunar Prospector 1998 Orbiter/Impact Success

Successful 21st Century Moon Missions

Smart-1 Europe 2003 Orbiter/Impact Success[32]
Selene Japan 2007 Orbiter/Impact   Success
Chang'e 1 China 2007 Orbiter/Impact Success[33]
Chandrayaan – 1 India 2008 Orbiter   Success[34]

---

[31] First Japan Moon mission.
[32] SMART-1 was used to test <u>solar</u> electric propulsion and other deep-space technologies, while performing scientific observations of the Moon. First European mission to the Moon.
[33] First China mission to the Moon.

Lunar Reconnaissance Orbiter USA 2009 Success[35]
LCROSS USA Impact 2009   Success
Chang'e 2 China 2010 Orbiter   Success
GRAIL USA 2011 Orbiter   Success[36]
LADEE USA 2013 Orbiter   Success[37]
Chang'e 3 China 2013 Lander   Success[38]
Chang'e 5 China 2014 Orbiter   Success
Queqiao China 2018 Orbiter   Success[39]
Chang'e 4 China 2018 Lander   Success[40]
Chandrayaan – 2 India 2019 Lander   Success

Tycho crater's central peak complex, shown here, is about 9.3 miles (15 km) wide, left to right (southeast to northwest in this view). Photograph and caption courtesy NASA.

---

[34] First India mission to the Moon.
[35] Remains an active mission to date.
[36] Gravity Recovery and Interior Laboratory. Mission consists of twin spacecraft.
[37] Lunar Atmosphere and Dust Environment Explorer.
[38] Active mission delivered Yutu rover to Moon surface.
[39] Lunar relay satellite.
[40] Delivered Yutu 2 rover to farside of the Moon.

As the above latest nation-state launches to the Moon shows, the race to explore the lunar surface is intensifying. The thought of unrestricted access to the Moon and its resources is alarming and could lead to both conflict and chaos. For an indication of how the future may turn out, if there is no regulation of access, look at all the space junk currently orbiting our planet by unrestricted satellite launches.

The Passive Seismic Experiment was the first seismometer placed on the Moon's surface. It allowed scientists to learn about the internal structure of the Moon. Photograph and caption courtesy NASA.

Interior of the Moon

The Moon is layered, a little like a layered cake. While it was in a molten state the heavier material such as iron and other metals sank to the interior of the Moon. The lighter materials such as rock and mineral fragments and glass of varying composition floated to the top and became the regolith, or Moon soil of the surface.

Like the Earth, at the center of the Moon is a core of metal, mostly comprised of iron and nickel but also lesser amounts of other metals. Around the inner core is a fluid outer core making the overall thickness of its core four hundred and ten miles thick. The core is about 20% the diameter of the Moon, whereas the Earth's core is approximately 50% of its diameter.

This small core may account for why there is very little volcanic activity evident on the Moon. Over decades and even centuries of viewing the Moon, occasional flashes of light are seen that indicate there is some activity.

The Aristarchus region of the Moon seems to be a particularly active volcanic area. So much so that Neil Armstrong mentioned seeing 'illumination' in that area during the Apollo 11 mission.

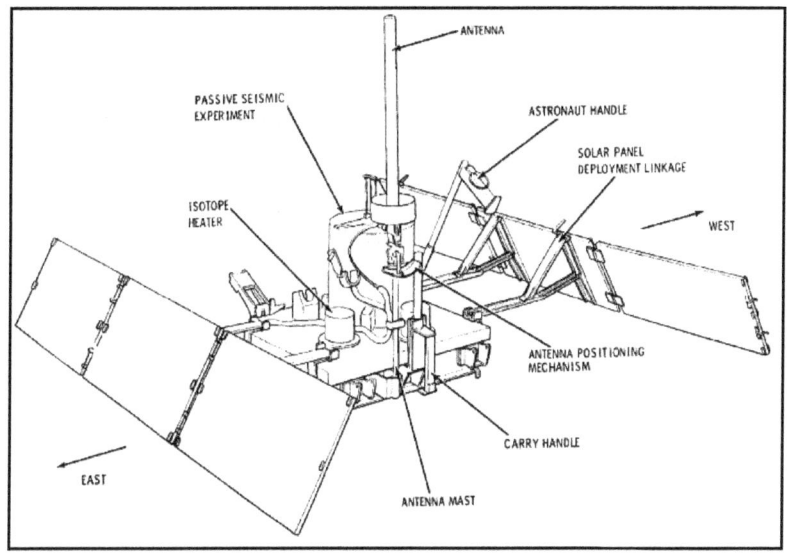

The Apollo 11 Passive Seismic Experiment device was left on the Moon surface. During its use on the Moon, this seismometer detected more than two hundred meteor impacts on the Moon. Drawing courtesy of NASA.

Layers above the core consist of the mantle and the crust. These layers are different in composition and once again indicate that the Moon was in a long term molten state in its early past. The mantle is much thicker at 839 miles on average as compared to the crust at 31 miles thick. As mentioned earlier in this narrative the crust of the far side of the Moon is about 30 miles thicker than the near side.

NASA, somewhat shyly, states that this difference in crust thickness is being researched as to why this may be. But you and I know why.

The Apollo astronauts, those hard working scientist-spacemen, left seismometers on the Moon.

These several devices reveal that the Moon experiences Moonquakes, some of them deep below the surface. It is thought these quakes are the result of Earth's gravity well and are, therefore, tidal events. As you know the Moon raises ocean tides on Earth and so Earth raises land tides on the Moon.

In effect the Earth's massive gravitational pull stretches the Moon as it orbits our planet. The heating and cooling of the Moon's surface also contributes to these quakes as it expands and contracts as the Moon moves from daytime to night and back again.

Other, currently unidentified forces also contribute to these quakes. I surmise these may be the result of volcanic activity inside the Moon.

Moon Mascons

Mass concentrations, are the remains of heavy-metal meteorites that long ago crashed into the moon and buried themselves. These massive objects can actually reach up with fingers of invisible gravity and tug on a spacecraft, sometimes violently.

One such mascon, a massive one, is located at the south pole of the Moon at the Aitken basin which is the largest intact impact crater known anywhere in the solar system. This mascon will have to be considered as we plan our return to the Moon in 2024 because the lander is slated to land in this same area.

Researchers seem to believe this is the remains of a large iron-nickel asteroid impact that struck the

Moon sometime in its deep past. It is believed to weigh in the neighborhood of 2.4 quadrillion[41] U.S. tons.

Latest research however, suggests that rather than a direct hit, a large space rock struck the Moon a glancing blow. Researchers suggest that the asteroid approached at a 30 degree angle resulting in a shallow strike deep enough to form a crater without any deep penetration. But it did leave enough of itself behind buried in the Moon to form a massive mascon.

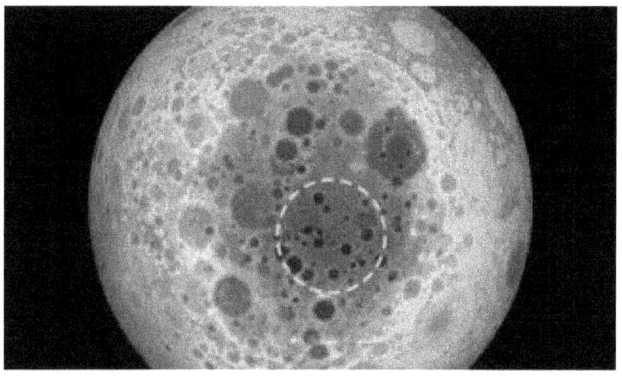

Location of the mascon at the Moon's South Pole. Image courtesy of NASA.

In August of 2019 it was announced that the first U.S. lunar lander in almost fifty years would launch in 2021.

NASA dubbed Astrobotic Technology Inc. for the unmanned lander mission to be launched by United Launch Alliance on its Vulcan rocket. This is

---

[41] Yes, that is an actual number and another way of saying it is a thousand raised to the power of 5.

the first of twenty missions to explore the lunar surface robotically over the next decade.

Our planet and its big moon appear to point the way for our search among exoplanets for an Earth II. It suggests that exoplanets that have big moons like ours and orbit in its star's habitable zone, is a good candidate for liquid water and an Earth-like planet.

Moon Magnetic Umbrellas

Scientists have recently noted at least one area of the Moon that may be shielded to some degree from the harsh solar wind. Termed a magnetosphere, this shielded area is 224 miles wide and could be a good site to establish a moon base.

Future research may discover more magnetospheres like this that could provide a magnetically protected safe harbor for human endeavors on the Moon.

The Moon, Our Stepping Stone to Mars

The Moon is, in my opinion, the only place we can construct spaceships to go to Mars and beyond with any efficiency. These ships by necessity will have to be very large, the largest we have ever built to hold the number of people we will want to send not to mention all the food, water and equipment needed for a long stay. And, it is my belief we will need to send ships in twos or maybe even threes and maybe even more backup ships, if we want to safeguard our traveler's safety.

The primary ship will carry our intrepid space explorers, the other two, just as large or even larger will be supply ships hauling food, water, and equipment necessary for survival on Mars. Why two? Because it's a long, long way back home should there be an emergency. Two ships would allow some room for error if say, for instance, we lost one on the way or during the attempt to land.

The landing upon Mars is likely the most dangerous portion of the trip because it offers the opportunity for the catastrophic loss of our people.

The trip to Mars cannot be simply a mission similar to a Moon mission it will have to be more in the realm of an expedition. We cannot rush this and we must do all we can to insure that when we depart for Mars, we have given ourselves every chance for success. Once we claim the Moon and learn the space lessons it will teach us, we will have plenty of time to plan our expansion into the solar system and do it as carefully as we possibly can.

The Moon is the place to build, equip, and launch such a major expedition. Assembling all the component parts and launching from the lunar surface with its one-sixth Earth gravity will be much less expensive than trying to leave from Earth and much more safe.

This is just a preview, a taste of what a Mars mission will entail. I will go into much more detail in a later book dedicated to this subject. But for now, we need to get back to the Moon.

Layered Regolith in the Von Karman Crater

The Chinese lander Chang'e 4 and its rover Yutu 2, have been on the farside of the Moon for more than a year. It touched down on the lunar surface near the South Pole on January 2, 2019. The rover is equipped with ground penetrating radar and records data as it is gathered. The results of the radar were published in the journal *Science Advances* on February 26, 2020.

Researchers poring over the information have detected three distinct and separate layers of material atop and beneath the lunar surface.

The first (top) layer consists of lunar regolith (Moon dust) studded with the occasional larger rock to a depth of 39 feet.

The second layer is varied in depth from 39 to 79 feet, and is composed of a coarser, heavier and larger-grained material as well as even more and larger embedded rocks. It is this layer that has the most rocks.

The deepest third layer that can be measured goes as far as 130 feet deep and has alternating bands of both fine and coarse grains. This layer also has larger rocks in it.

The embedded rocks are likely ejecta[42] deposits from crater impacts. Meanwhile, like our Mars rovers, Yutu 2 seems to continue to operate far beyond its expected perimeters.

---

[42] Material that is forced or thrown out, especially as a result of volcanic eruption, meteoritic impact, or stellar explosion.

CHAPTER 2
We Return to the Moon

Our Biggest Past Moon Error

The very successful Apollo Moon missions ended on December 19, 1972 when astronauts Eugene Cernan, Harrison Schmitt, and Ronald Evans splashed down in the Pacific Ocean.

At the time I was an Army sergeant serving with the combat engineers in Germany. Like millions of other Americans including the space workers and scientists of NASA, I was greatly disappointed that we had deserted the Moon and any chance of remaining there by establishing permanent bases.

For years after the end of Apollo I watched with great interest the advent of our space shuttle program. And yet, I could not but continue to lament what I saw as a major error. We had settled for something far less than we could have attained.

But I was busy with my military career and taking care of my family. I felt voiceless, like many other citizens to have any impact on what our short sighted politicians were doing to NASA and our space program.

Now, thanks to the far sighted scientists and administrators at NASA and a president who is very enthusiastic about returning to the Moon and eventually going to Mars, I feel we are back on the right track to insure our future.

During those many years I completed my military career, and embarked on other endeavors including obtaining an advanced degree and becoming a writer. After the successful publication of many books, I decided to voice my opinion and become an active advocate of our space program.

The Moon is extremely important to human survival. We must become a two world species in order to avert the destruction of all humankind in some planetary catastrophe.

Many tout our proposed mission to send humans to Mars. But the truth of the matter is that going to Mars and surviving there is a very hard and very serious gamble. I do believe we will get there but certainly not in the near future.

The Moon is much closer, in an emergency we can get a spacecraft there in a matter of days instead of the months it would take to reach Mars. In my view the Moon is an essential stepping stone in learning to live in a hostile environment. The lessons we learn on the Moon will pave the way to the exploration of the rest of the solar system. Without taking this more traveled road, we cannot be prepared not to make a second major error in our endeavor to explore outer space.

In the end, the Moon may be the only other celestial body we will ever live on. This is something we must do and we must do it now. And on the way, we must insure it does not fall under the control of some repressive form of government.

NASA's Artemis Program

Paraphrased below is the NASA explanation for the name of the new manned Moon Program. Of course everyone remembers the name of our first manned program to the Moon was Apollo.

Artemis is the twin sister of Apollo and goddess of the Moon in Greek mythology. Now, she personifies our path to the Moon as the name of NASA's program to return astronauts to the lunar surface by 2024, including the first woman and the next man. When they land, our American astronauts will step foot where no human has ever been before: the Moon's South Pole.[43]

Working with U.S. companies and international partners, NASA will push the boundaries of human exploration forward to the Moon for this program. As a result of Artemis, NASA will be able to establish a sustainable human presence on the Moon by 2028 to uncover new scientific discoveries, demonstrate new technological advancements, and lay the foundation for private companies to build a lunar economy.

With our goal of sending humans to Mars, Artemis is the first step to begin this next era of exploration.

The Artemis Space Program's mission is to return American astronauts to the moon, with the goal of landing 'the first woman and the next man' on the lunar South Pole region by 2024. Landing at the pole is a major event twice over, the significance of which will become clear in the following pages of this narrative.

Artemis is seen as the first attempt toward a permanent US presence on the Moon and the establishment of commercial enterprises and a

---

[43] The South Pole is one of the locations of water on the Moon.

lucrative lunar economy that will pay great financial dividends.

The Moon is a treasure chest of both science and money, money in the trillions of dollars.

Space Policy Directive 1

In the month of December 2017, on the 45th anniversary of the last manned mission to the lunar surface, President Donald J. Trump approved a directive that includes a lunar mission on the pathway to Mars and beyond.

We'll learn. The directive I'm signing today will refocus America's space program on human exploration and discovery. It marks an important step in returning American astronauts to the Moon for the first time since 1972 for long-term exploration and use. This time, we will not only plant our flag and leave our footprint; we will establish a foundation for an eventual mission to Mars. And perhaps, someday, to many worlds beyond.

— President Donald Trump, 2017[1]

White House Moon Spending Plan 2021

Fiscal year 2021 begins in October of 2020 and contains a $25.2 billion budget for NASA. Although

I think we need an even more generous increase in view of China's push to the Moon, it does amount to a 12% increase in funding. Approximately half of this money will go toward the vital program to return our astronauts to the Moon. More than 3 billion dollars of the budget will go toward the development of a lunar lander capable of delivering the first woman and next man on our Moon.

Other NASA developments receiving money includes the next generation of spacesuits, lunar surface habitation construction, and Moon rovers. It is a step in the right direction as we move forward toward the goal of a 2024 manned Moon landing.

NASA Administrator Jim Bridenstine's Statement on the Moon to Mars Initiative, and the Fiscal Year 2021 NASA Funding

"President Donald Trump's Fiscal Year 2021 budget for NASA is worthy of 21st century exploration and discovery. The President's budget invests more than $25 billion in NASA to fortify our innovative human space exploration program while maintaining strong support for our agency's full suite of science, aeronautics, and technology work.

The budget proposed represents a 12 percent increase and makes this one of the strongest budgets in NASA history. The reinforced support from the President comes at a critical time as we lay the foundations for landing the first woman and the next man on the South Pole of the Moon by 2024. This budget keeps us firmly on that path.

We are preparing to achieve pivotal milestones this year in development of the Space Launch System rocket, Orion spacecraft, and the Gateway. These make up the backbone of our Artemis program and are fully supported by this budget. They constitute our ability to build a sustainable lunar presence and eventually send human missions to Mars.

Most noteworthy, is the President's direct funding of more than $3 billon for the development of a human landing system. This is the first time we have had direct funding for a human lander since the Apollo Program. We are serious about our 2024 goals, and the President's budget supports our efforts to get the job done.

We soon will launch American astronauts on American rockets from American soil for the first time in nearly a decade. This recaptured ability will not only allow us to do more science and more exploration than ever before, but will also broaden commercial activity in low-Earth orbit to support ever greater private partnerships.

As we prepare to celebrate 20 years of continuous human presence aboard the International Space Station this year, we will continue to look for ways to partner with private enterprise and give more people access to the unique environment microgravity offers. Similarly, when we go to the Moon in the next four years, we are interested in taking the world with us. This includes those involved in our Commercial Lunar Payload Services initiative and the international relationships we have forged over the decades.

The FY 2021 budget positions NASA to spearhead a new era of human space exploration without focusing funds on one program at the expense of others. This all-of-NASA approach to the future will help us take advantage of all the exciting, new horizons emerging in science, aeronautics, and technology.

The decadal survey priorities are strongly supported by this budget; including history's first Mars sample return mission, the Europa Clipper, and development of a host of new trailblazing Earth observation missions. In aeronautics, the budget backs all our cutting-edge research on commercial use of supersonic aircraft, all-electric airplanes, and development of an unmanned aerial system that will make flying small drones safer and more efficient in the 21st century.

NASA is on the cusp of embarking on era-defining exploration. The civilization-changing technology we develop will deepen humanity's scientific knowledge of the universe and how to take care of our ever changing world.

I am confident the FY 2021 budget's proper investment in our agency's priorities, coupled with your unmatched talents and expertise, will strengthen our national posture for continued space preeminence and, as President Trump said during his State of the Union speech last week, help our nation embrace the next frontier."

NASA's new Gateway logo. The colors you cannot see are red, white, and blue, with the distant planet red, for Mars. Courtesy, NASA.

Orion Spacecraft

The Orion Multi-Purpose Crew Vehicle (Orion MPCV is a combined U.S. and European spacecraft intended to carry a crew of four astronauts to destination at or beyond low Earth orbit.

Currently the spacecraft is still under development by NASA and the ESA for launch on

NASA's Space Launch System. Orion is intended to be the main crew vehicle of the Artemis lunar exploration program as well as future crewed flights to Mars and possibly other destinations.

Artemis 1 will be the first flight of the Orion on the Space Launch System. The second mission, Artemis 2 will be the first crewed flight, and Artemis 3 the first lunar landing via the Lunar Orbital Platform known as Gateway.

NASA's Marshall Space Flight Center in Alabama has been selected to develop a lunar lander for Artemis. According to NASA Administrator Jim Bridenstine, the center was selected to be in charge of the lunar lander because of its extensive experience with propulsion systems, which will be crucial for executing safe landings and liftoffs from the lunar surface. Bridenstine also stated that Lisa Watson-Morgan, a long term employee at Marshall will oversee the lander program.

You might be wondering what the Space Launch System and Lunar Orbital Platform - Gateway is. That's understandable. These are new concepts for NASA.

The Space Launch System refers to the rocket that will carry Orion into space. This heavy lift rocket is still under development and is expected to carry Orion's crew to the Moon and later to Mars.

The gateway is a space station designed to be placed in orbit around the Moon and serve as a space science laboratory, a communications hub and short term habitat for space farers. It can also hold supplies and equipment for eventual shipping to the lunar surface.

The project is supervised by NASA and meant to be a gateway for civil commercial and international partners. It will serve as the staging area for both crewed and robotic exploration of the Moon's South Pole, and is the proposed stage for NASA's Deep Space Transport system.

A NASA artist's concept of the Space Launch System.

Recently, storm clouds have arisen on the horizon of the Space Launch System. Boeing, the contracted builder of the powerful rocket is behind schedule by two years and experiencing cost overruns. If it launches on its first flight in 2021, the system will have already cost $18 billion dollars, two billion more than expected. The second launch carrying astronauts in 2023 will balloon the costs to $23 billion dollars. This may trigger a hard look at the program by congress.

NASA's Office of the Inspector General placed the blame on "…management, technical and infrastructure issues driven by Boeing's poor performance."

The OIG's report recommended five steps to improve NASA's management of costs and compliance with congressional directives for any portion of the program that exceeds its projected funding by 30% or more.

NASA's response was they would implement the new procedures.

The cost overruns were attributed to technical problems in actually building the rocket and its engine control systems.

The SLS core stage intended for the first Artemis mission in 2021 is hauled to a waiting barge for a January trip to the Stennis Space Center in Mississippi. Mounted in a massive test stand at Stennis, engineers plan months of work to check out the rocket's systems before a full-duration main engine test firing later this year. NASA.

Funding Artemis

NASA has asked for what they refer to as a down payment of 1.6 billion dollars to return our astronauts to the Moon by 2024 and establish a permanent Moon base by 2028.

In my opinion they should push to do both at once. When we get to the Moon in 2024 we need to start building a base immediately. In fact it would be much more politically desirable to get to the Moon even earlier, say by 2022. I think if we put enough resources into doing this that 2022 is a reasonable goal not withstanding Boeing's problems with the Space Launch System.

I feel this way because assuming President Trump gets a second term in 2020 there will of course be a presidential change by November 2024. And that likely means a change in the political will needed for us to continue pushing forward on the Moon and Mars. If we falter then, you can believe that the Communist Chinese will pick up the mantle with a vengeance and our claim of primacy over the Moon will mean little if anything. Remember that the Chinese Communists don't need to rely on a willing constituency of their people to do what they like.

Weak kneed members of congress are already voicing their skepticism and concerns about NASA's plans. Without strong presidential leadership this opportunity to begin the true expansion of humans from Earth, at least by a freedom loving nation, will be lost and probably forever.

Recently, NASA administrator Jim Bridenstine said "Given our current rate of production, we will have three SLS's available by 2024 and the third one would be for Artemis 3, that takes us to the moon in 2024. I think that is fully within the realm of possibility, but a lot of things have to go right to make that happen."

Indeed they do, both funding wise and politically.

An artist's concept of the Lunar Orbital Platform – Gateway image courtesy, NASA.

Moon Nuclear Power

In August of 2019 President Donald Trump signed an executive memorandum authorizing NASA to use nuclear powered propulsion in outer space. This is a very big deal and a real boost to the exploration of space.

The memorandum is comprehensive and likely much more in depth than the reader would want to read. For that reason I have restricted it to the opening phrases as follows:

SUBJECT: Launch of Spacecraft Containing Space Nuclear Systems

By the authority vested in me as President by the Constitution and the laws of the United States of America, I hereby direct the following:

Section 1. Purpose. This memorandum updates the process for launches of spacecraft containing space nuclear systems. Space nuclear systems include radioisotope power systems (RPSs), such as radioisotope thermoelectric generators (RTGs) and radioisotope heater units (RHUs), and fission reactors used for power and propulsion…

And, of course, signed by the president.

Kilopower Project

The Kilopower Project is a joint endeavor by NASA and the U.S. Department of Energy to put nuclear power plants to work in space, on the Moon, and on Mars. This project is already in the works and the signing of the executive order by the president is right on time.

A NASA statement from their Space Technology Mission Directorate (STMD) reads:

"The Kilopower project is a near-term technology effort to develop preliminary concepts and technologies that could be used for an affordable fission nuclear power system to enable long-duration stays on planetary surfaces."

Basically this means that while solar power and other sources of energy may be and likely will be used, long periods of staying the Moon and eventually Mars, will require nuclear fission power.

This will meet the energy requirements of any kind of settlement whether it's a commercial outpost, a colony of any size, a city inside a natural cavern, or a Space Force base.

The Kilopower reactor passed its recent ground tests with no noted problems. The confident project leader Patrick McClure says that not only is the project on its way to reality, it will be ready in the very near future. He added: "I think we could do this (complete the reactor) in three years and be ready for [space] flight."

MTSD added "The Kilopower project team is developing mission concepts and performing additional risk reduction activities to prepare for a possible future [space] flight demonstration."

Nuclear power in our space endeavors is not new. NASA's own Voyager 1 and Voyager 2 spacecraft, the New Horizons spacecraft that traveled to Pluto and beyond, and the Curiosity Mars rover, along with other robot explorers, use radioisotope thermoelectric generators (RTGs) that convert the heat generated during the radioactive decay of plutonium-238 into useable electricity.

The Kilopower Project will pave the way for groundbreaking human endeavors that will unlock the science and bounty space exploration and open the Moon as our gateway to the solar system.

Nuclear Propulsion Spacecraft

An illustration of a spacecraft powered by nuclear thermal propulsion in orbit. Image courtesy NASA.

Nuclear powered spacecraft are so much faster that one could travel to Mars in half the time it takes current spacecraft.

Spacecraft powered by such engines could conceivably reach Mars in just three to four months, approximately half the time of the fastest possible trip in a vehicle with traditional chemical propulsion, according to NSC panelist Rex Geveden, the president and CEO of BWX Technologies Inc.

That change of speed should translate into a big financial savings not only for NASA, but for future commercial flights to Mars as well.

Some have said it's a very big deal, and I agree. Director Jim Bridenstein stated "That is absolutely a game-changer for what NASA is trying to achieve. That gives us an opportunity to really protect life, when we talk about the radiation dose when we travel between Earth and Mars."

On Mars, illustration courtesy NASA.

Nuclear Cislunar Rocket

To extend our satellites further out from Earth and deep into the cislunar[44] area of space we have turned to the development of a nuclear rocket. These would take the form of maneuverable satellites located thousands of miles above the low Earth orbit satellites of today.

---

[44] Cislunar refers to that area of space between the Earth and the Moon.

The United States and China and perhaps Russia are in competition for using the cislunar area to advantage.

A written proposal by DARPA (Defense Advanced Research Projects Agency) stated:

"The capability afforded by nuclear thermal propulsion will expand the operating presence of the U.S. in space to the cislunar volume [area] and enhance domestic operations to a new high-ground, which is in danger of being defined by the adversary, [such as China or Russia]".

"An agile nuclear thermal propulsion vehicle enables the Defense Department to maintain space domain awareness of the burgeoning activity within this vast volume." Jared Adams, a DARPA spokesperson said.

Because of the danger inherent in nuclear propulsion it is envisioned these atomic powered rockets would be assembled and launched in space. This would require specially trained astronauts and involve some danger for them.

NASA illustration of a nuclear Moon rocket.

Cosmic Radiation

Cosmic radiation, sometimes called cosmic rays, are a kind of high energy composed of protons and atomic nuclei originating from the Sun and from beyond the solar system. A significant fraction of them come from supernova explosions of stars.

This radiation is dangerous to humans and in massive amounts, is lethal. Cutting down on the travel time to Mars could be the difference between life or death by radiation poisoning. So obviously nuclear propulsion is a good thing, albeit not without its own inherit risks.

High value satellites such as our GPS system equipped with propulsion capability and added detection radars could conceivably avoid any errant threat of collision or intended attack.

The Moon, unlike Mars, is close by the Earth, it can be reached in a matter of days and eventually in hours by spacecraft. That's important, it means that if our people on the Moon suddenly need succor from an emergency, we can respond quickly enough to avert a catastrophe.

New Spacesuit

Although NASA is researching a new spacesuit dubbed the xEMU for Exploration Vehicular Mobility Unit, very much like a one person spaceship.

In September of 2019 after a meeting of the independent NASA Aerospace Safety Advisory Panel (ASAP) the panel recommended that NASA begin a new spacesuit program. This would be in addition to the xEMU and concentrate on a suit for use on the lunar surface.

The panel pointed out that the only operational EVA (Extra Vehicle Activity) suits are currently on the ISS and they are forty years old.

A recent rash of damages to astronaut's gloves while in the EVA suits point out a glaring weakness. A metal railing around the exterior of the space station intended for use of the astronaut's while doing their tasks has been impacted by micro meteorites. These impacts created tiny craters in the railing with jagged edges that cut the gloves of the EVA suit. This poses a danger to the astronauts and could pose the possibly lethal risk of suit decompression.

This risk is the reason for the cancellation of the first woman to go EVA scheduled earlier this year.

The panel recommended a new spacesuit that would be appropriate for use on the lunar surface.

A prototype of the xEMU space suit. Courtesy, NASA.

The xEMU is still in development but its basic features are mostly complete at this writing. Although space testing isn't scheduled until 2023 a seemingly unnecessary long time from now, it is undergoing testing underwater. It would seem to me that a working prototype could be assembled in the next year (2020) and flown up to ISS for unmanned testing in space. And, we already have a good working background to move forward on, we are not starting from scratch. A good example of what can be done in a timely manner was the embarrassing fact that NASA did not have enough medium sized spacesuits to conduct an all-female spacewalk. That problem was quickly dealt with by making new spacewalk spacesuits and quickly. I bet it can be done with the xEMU too. Come on guys, time, as they say, is money!

NASA must have been listening, because they didn't waste any time. At 7:38 am on October 18, 2019 astronauts Christina Koch and Jessica Meir exited the ISS for the first all-female spacewalk. Their mission was to repair a faulty electrical power system. In a conversation with President Donald Trump by radio Meir said

"...we recognize that it (the spacewalk) is a historic achievement, and we do of course want to give credit to all those that came before us," she told the president. "There has been a long line of female scientists, explorers, engineers and astronauts, and we have followed in their footsteps to get us where we are today. We hope we can provide an inspiration to everybody — not only women, but to everybody that has a dream, that has a big dream, and is willing to work hard to make that dream come true."

Koch is a veteran spacewalker, but this was the first for Meir who is new to space. The repair went without a hitch and the two women remained outside for an extra hour in their new spacesuits.

Lunar Landers

This NASA illustration depicts an artist's concept of a lunar lander module lifting off from the Moon to return to orbit.

In September 2019 the space agency called for the design and eventual development of a new human capable Moon lander. The obvious downside of this is that starting from scratch is going to take a long time, likely as long as six to eight years! This seems redundant as we already know how to make lunar landers, we've done it before.

Instead of starting over, why not gather the best engineers from all sources and using what we already know, get a faster product designed and completed? This would require cooperation from the best and brightest of the space commercial companies.

To seemingly aggravate the situation, concerns voiced by the commercial companies resulted in NASA dropping the requirement that the landers be able to be refueled. That seems like a critical requirement to just let go.

Companies like Blue Origin, SpaceX, and Lockheed Martin are already working on Moon landers. I would think that the first one to build a model capable of being refueled would get the NASA funding and corner the future of human capable space landers.

In November of 2019 NASA invited the usual space companies of SpaceX and Blue Origins, as well as three lesser known outfits to submit their lunar lander ideas.[45] These five companies can now join NASA's Commercial Lunar Payload Services program (CLPS). One or more of them will be selected to deliver payloads to the lunar surface. These additions bring the number of NASA recognized commercial companies working on lunar landers to fourteen.

The CLPS program is intended to bring the innovation and expertise of American industrial might to get to the Moon quickly.

This past July[46] NASA awarded the first three government contracts awarding lander missions to Astrobotics, Intuitive Machines, and Orbit Beyond.

---

[45] Including Tyvak Nano-Satellite Systems, Ceres Robotics, and the Sierra Nevada Corp.
[46] 2019

Astrobotics and Intuitive Machines payloads will consist of science and technology experiments and will launch this year (2020).

The Intuitive Machines Nova-C lander mission will ride a SpaceX Falcon 9 rocket, while Astrobotic's Peregrine lander will fly aboard a United Launch Alliance rocket.

NASA has a total of fourteen payload contractors that may potentially deliver loads in its CLPS program. Its NASA's intent to send two missions a year initially to the Moon.

NASA expects to decide on up to four companies to study concepts and build prototypes for a human lander in early 2020.

Blue Origin, Lockheed, Northrop, Boeing, and Sierra Nevada Corp. partnered with Dynetics Inc. are all expected to submit bids to build the human landers.

Initially NASA will award several contracts for the bidding companies to complete design concepts on their landing systems, after which it will decide upon a contractor to build a lander. The winning contractor will get a contract likely worth in the billions of dollars.

Walking and other Moon Rovers

There are a lot of niches in the coming and ongoing exploration of space for those with an active imagination. A United Kingdom company with the rather cute name of Spacebit plans to send a very small walking rover to the Moon by hitching a ride on a rocket going their way. Spacebit recently signed a contract with U.S. Astrobotics to ferry the rover on their Peregrine lunar lander scheduled to launch in 2021.

The rover's mission is to enter and explore a lunar lava tube. This will be the first attempt to explore this Moon feature and determine its environment and possible suitability for human habitation. Speculation is that lava tubes may be sealable and able to hold a breathable oxygen atmosphere. It is also thought the rock walls could provide protection from solar and cosmic radiation as well as possible impacting meteors.

The miniscule arachnid-like rover is expected to only last one day of exploration. This obviously short period of time is thought to be enough to determine the feasibility of deploying swarms of exploring rovers in later missions. The data collected for use by scientists and engineers alike may prove invaluable to future bases and settlements.

So, put your thinking caps on, you could be the next space entrepreneur to think of a space exploration niche your future company can take advantage of. And if you do think of something, what do you do first? You take your idea to NASA's Innovative Advanced Concept (NIAC) program. NIAC funds innovative ideas for new kinds of telescopes, propulsion systems, and other space exploration ideas. If they find your concept worthy of pursuit, they could fund you to the tune of a couple hundred thousand dollars so you can test its feasibility.

As they did for Moon landers in September, NASA is asking for input from space companies and others that may not be in the business, on the subject of Moon rovers in February. This time they are also soliciting ideas from other tech companies and vehicle manufacturers.

One of the things they are seeking in their request for information (RFI) is ideas for a human Moon car something like the Moon buggy of the Apollo landings. A new Moon buggy would conceivably have the latest in battery propulsion and the latest in computer software and so would have a big advantage over the 1970s model. It would also likely be more impervious to the insidious invasion of Moon dust.

However, ideas for the less glamorous robotic rovers are also sought. The intent of the rovers is to give the astronauts the ability to explore a larger area around their landing site. The first Moon landing mission is aimed at the Moon's South Pole where water is likely nearby in the form of ice. The robotic rovers would be able to go further and deeper into recesses than NASA may want to send the astronauts.

NASA has a somewhat fast closing date for RFI submissions with the human rated rover due in February and the robotic rover submissions by March 6, 2020. This would seem to suggest NASA already has an indication that a number of companies are interested and perhaps already equipped to respond to the rover challenge.

Project Farside

The farside of the Moon is thought to be a good place to study the universe. The farside permanently faces away from Earth and so would be virtually free from straying radio and TV transmissions from Earth. This allows the use of a radio telescopic without interference from all the constant broadcasting on our planet. That and other considerations gave birth to the FARSIDE project, an acronym for 'Farside Array for Radio Science Investigations of the Dark ages and Exoplanets' by interested scientists who prepared the mission concept for submission to NASA.

The mission would permit around the clock observation of nearby star systems, and exoplanets for characteristics indicating habitability. It would also allow observation of the solar system and deliver images that will rival those of space telescopes.

Other tasks could include ground penetrating radar to sound the lunar subsurface for such things as water and mineral deposits.

The FARSIDE concept, besides the instruments themselves, includes a Moon lander, base station for personnel and a rover for deploying equipment and antennas to cover a wide area of the Moon surface. Transmission of data would have to be sent initially to a relay satellite for retransmission to Earth.

The project team estimates the cost of the FARSIDE project at an initial 1.3 billion dollars.

Illustration courtesy NASA.

Chapter 3
First Commercial Astronauts

On the 8th day of August, 2018 NASA announced the names of the first commercial astronauts scheduled to fly missions on both the SpaceX Dragon and Boeing Starliner capsules. SpaceX was to test launch their space vehicle in November of 2018 while Boeing was scheduled for some time later.

U.S. astronauts have not flown aboard an American spacecraft since the Space Shuttle program ended in 2011.

The astronauts were presented to the press and the public by NASA administrator Jim Bridenstine. Three of the new crewmen have not flown in space before.

Victor Glover and Mike Hopkins of SpaceX (NASA)

Bob Behnken and Doug Hurley of SpaceX (NASA)

Eric Boe, Nicole Mann, Chris Ferguson of Boeing (NASA)

John Cassada and Suni Williams of Boeing (NASA)

Astronaut's Background

Bob Behnken of SpaceX is an Air Force colonel and has flown in space twice aboard the shuttle *Endeavor* and has been an astronaut since 2000. He has a total time in space of approximately thirty days including nineteen hours of space walking. Behnken has a Ph.D in mechanical engineering, and flew aboard space shuttles as a mission specialist. He was formerly the Chief of the Astronaut Office.

Behnken is married to Katherine Megan McArthur who is also a NASA astronaut and was one of the NASA crew that serviced the Hubble Telescope.

Bob Behnken and Megan McArthur (NASA)

Eric Boe of Boeing is a former U.S. Air Force colonel and test pilot, and has also been an astronaut since 2000 piloting two space shuttle missions. Boe has a science degree in astronautical engineering and a master degree in electrical engineering. Boe is the former Deputy Chief of the Astronaut Office. He is married to Kristen (Newman) Boe and they have two children.

John Cassada of Boeing is a Navy commander and a test pilot. He has been an astronaut since 2013 but has not yet been to space. John holds advanced degrees and is trained as a physicist.

Eric Boe (NASA)

John Cassada (NASA)

Chris Ferguson of Boeing is no longer an active NASA astronaut; however, he did pilot three space shuttle missions. He resigned from the astronaut corps after the shuttle program was terminated. Ferguson is the director of crew and mission operations for Boeing. He holds degrees in mechanical and aeronautical engineering.

Chris Ferguson serving as CAPCOM for a Space Shuttle Mission (NASA)

Victor Glover of SpaceX is a Navy Commander and test pilot and was also selected for the astronaut corps in 2013. His first trip to space will be aboard the Dragon's first full mission flight.

Victor J. Glover (NASA)

Mike Hopkins of SpaceX became an astronaut in 2009, he is an Air Force colonel, an experienced flight test engineer, and has been to space twice on shuttle missions. Hopkins earned a degree in aerospace engineering from the University of Illinois at Urbana where he played defensive back for the Fighting Illini football team. In 1992 he obtained an advanced degree in that same discipline from

Stanford University. Mike is married and has two sons.

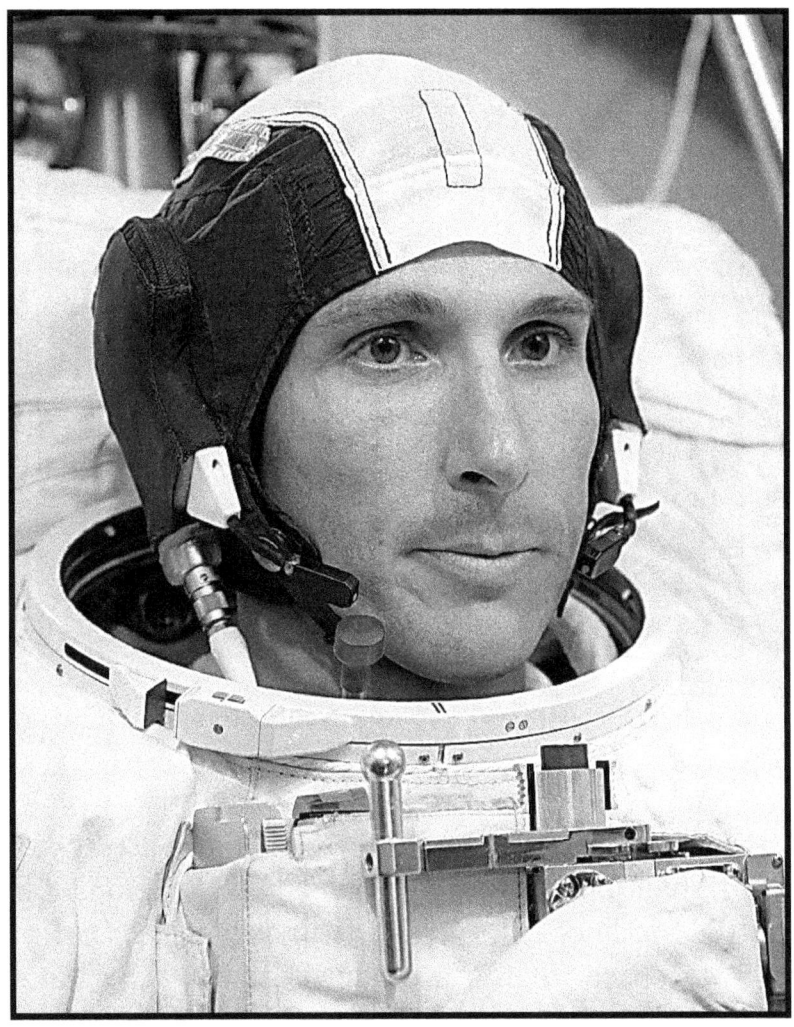

Mike Hopkins (NASA)

Doug Hurley of SpaceX joined the astronaut corps in 2000 and is a former Marine Corps test pilot. He piloted two of the space shuttle missions, including the final space shuttle flight in 2011. Doug's call sign is 'Chunky', he is married to fellow astronaut Karen Nyberg.

Doug Hurley and Karen Nyberg (NASA)

Nichol Mann of Boeing, one of the two women astronauts, was a test pilot of the Marine Corp's F-18 fighter jets before becoming a NASA astronaut in 2013. Starliner's first crewed flight will be her first trip to space. She is a graduate of the U.S. Naval Academy and holds a degree in mechanical engineering from Stanford University.

Nichol Mann (NASA)

Sunita (Suni) Williams; the other woman astronaut, has been an astronaut since 1998. She is a former Navy captain and test pilot. Suni has accumulated a year of space duty and has performed seven space walks. She has been to the International Space Station on expeditions 14 and 15 then on expeditions 32 and 33 and was in command of expedition 33.

Sunita (Suni) Williams (NASA)

The First Corps of Commercial Astronauts (NASA)

This is a group of both star studded and potential future star astronauts. It is clear that they have been carefully picked to lead the United States and indeed the world into a new space endeavor during which they will boldly go where no one has gone before!

Women in Space and On the Moon

In its introduction to our intended return to the Moon NASA mentions it is the return of the next man and the first woman to the Moon. It should be obvious to anyone who has an interest in space exploration that women most assuredly do have a place in our space program. It certainly is obvious to me. I have noted with admiration the dawn of the era of women astronauts.

Dr. Nadia Drake, a science journalist, wrote an article titled *Let's Send Only Women to Space* for National Geographic. When I saw the title I thought to myself "Well, that sounds a little selfish to me but I'll read her opinion."

Dr. Drake asserts that women have the right stuff physically and psychologically for lengthy missions in space. I agree, however, I think men do too.

She then asks why we should send men along at all and then mentions a number of what I think of as diminutive factors. The first being the relative size of women as opposed to that of men. Yes, I agree that size matters in some areas such as cost. It would take less consumables to sustain a crew of smaller sized females as compared to a crew of more robust men.

However, I will point out that there are small men out there too. During WWII smaller men were well suited for the tight quarters in armored tanks. During the Vietnam War the smaller man was ideal as a 'tunnel rat', those individuals who could more easily travel down enemy tunnels dug for the diminutive size of the Vietnamese soldier. This seems to be a small concern (no pun intended) at least in missions inside the solar system.

The second factor mentioned is the seemingly less physical effects noted for women than men from long stints in space. It appears to me that this question is still out for determination. We will have to wait for more data on long term space exposure on women, for more definitive comparisons with information already to hand on men astronauts.

The third mentioned consideration is that women have some personality traits more suited for long term duration space missions. I think what I said regarding the second factor applies equally with this assertion.

The fourth consideration Dr. Drake spoke of, populating another world, and whether taking men along is necessary when simply taking a sperm bank would do, doesn't concern us on our short trip to the Moon.

In the end Dr. Drake is most correct when she asks "When will there be enough women in the spacecraft? When everyone who's qualified has an equal shot at a seat." I certainly agree.

Astronaut Jeanette Epps, one of the women under consideration to go to the Moon. (NASA)

The new astronaut graduates are Kayla Barron, Zena Cardman, Raja Chari, Mathew Dominick, Bob Hines, Warren Hoburg, Jonny Kim, Jasmin Moghbeli, Loral O'Hara, Frank Rubio, Jessica Watkins as well as Canadian Space Agency graduates Joshua Kutryk, and Jennifer Sidney-Gibbons. Great job! NASA photo.

First Generation of Artemis Astronauts

On January 10 at the Johnson Space Center in Houston, NASA graduated its first Artemis generation of astronauts consisting of six women and seven men. Two of them are Canadians.

These deserving thirteen people were selected from more than eighteen thousand applicants and completed a two year basic astronaut training program to qualify for graduation and receive their astronaut pins.

They come from backgrounds that include pilots, engineers, scientists and doctors. As astronauts they join the long line veterans who have gone to space before them. They will help support the other astronauts in space and help develop spacecraft of the future. They will likely be among those who will eventually go back to Moon beginning in 2024.

NASA has already started recruiting the next class of astronauts and is accepting applications in March from those who are pilots and have an advanced degree in a hard science.

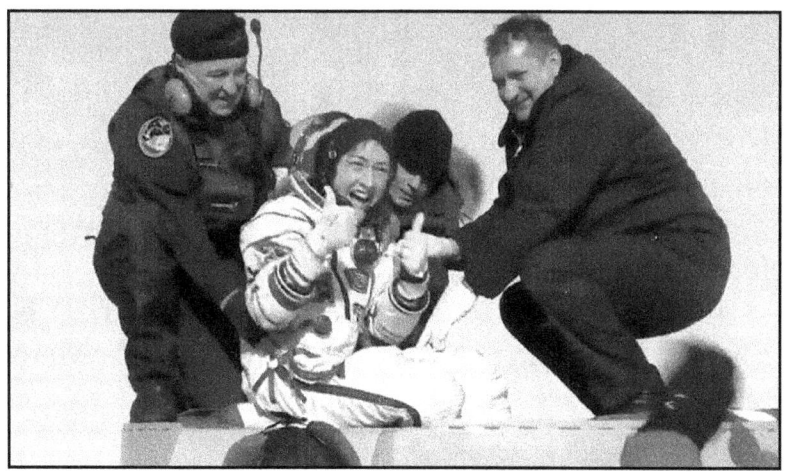

Christa Koch returns from space after her record setting flight. NASA photo.

Christa Koch Sets Space Flight Record

Right at the end of 2019 on December 28th, forty year old Christa Koch set a new record for the longest spaceflight by a woman astronaut. On that day she broke the old record of 288 days with a little over two months left in her current flight mission. The previous record was set by former space station commander Peggy Whitson during her space flight that spanned a period during the years of 2016 and 2017.

Christa is scheduled to spend a total of 328 days, nearly 11 months on board the ISS. The current record for all astronauts is held by Scott Kelly who stayed for 340 days during the years 2015 to 1016. However, the world record is held by a cosmonaut who stayed 15 months aboard the Mir space station.

Forty-one days after setting the previous record in space for a woman, on February 6, 2020 Christina Koch returned to Earth having established a new record in space of 328 consecutive days!

Cosmonaut Valeri Polyakov has the overall record of 437 days.

It's this kind of effort that could help Christa be selected as the first woman to step foot on the Moon or even on Mars.

Boeing's Starliner sits atop a United Launch Alliance Atlas V rocket in preparation for launch to the International Space Station (ISS).

Space Flight Test of the Starliner

The Starliner was test launched on December 20, 2019 and was supposed to fly to the International Space Station.

The unmanned launch itself was nearly flawless, but what was described as an anomaly in the Mission Elapsed Time, or basically the ship's clock, caused the burning of too much fuel. The upshot was that the Starliner lacked enough fuel after attaining low orbit that it did not have enough for the rest of the journey. This was a major failure but NASA pointed out that had astronauts been aboard the ship, the crew could have corrected the error and flown on to the ISS.

NASA illustration of the Boeing Starliner in orbit.

Never the less the Starliner was expected to remain in orbit for two as scheduled. The capsule will deploy braking parachutes and land in the New Mexico desert.

This mission, dubbed the "Orbital Flight Test' and its landing are crucial steps in our return to manned space flight. The OFT is a critical step in evaluating the Starliner. The intent now is to launch a three man crew in the ship sometime in 2020.

NASA picture of Boeing's Starliner after it safely landed in the desert of New Mexico, December 2019.

Blue Origin's New Shepard

A handful of days before Boeing's launch, on December 15, Blue Origin launched its New Shepard spacecraft. Its flight was suborbital and not meant to attain orbit. The launch was primarily to test the ship's ability to perform a powered landing, which it did without a hitch. The twelfth such launch and landing, it is another step in the mission of clearing the spacecraft for human flight.

The 10:30am launch of this SpaceX Falcon 9 rocket and successful crew escape test has paved the way to renewed manned space flight. Photo courtesy NASA.

SpaceX Crew Dragon Test Success

On January 19, 2020 SpaceX demonstrated its high level of mastery of space operations and paved the way for the U.S. return to manned space flight this year.

As the rocket attained its test altitude the Crew Dragon capsule launched from the spacecraft firing its own thrusters and sped away.

Behind and below the Crew Dragon the Falcon 9 dramatically exploded in midair as it was designed to do, simulating an emergency ejection of the crew.

A few minutes later the capsule deployed its drogue parachutes to slow the descent of the capsule and reduce the speed of its fall before the main parachute opening.

The next step was completed flawlessly as the four huge 'chutes blossomed in the sky to the roar of the watching crowd back on Earth. From that point the Dragon floated gently down to land in the Atlantic Ocean. Had this been an actual emergency the crew inside would have been completely safe. The demonstration was an unqualified success!

At the news conference following the test flight both SpaceX and NASA were obviously elated at the success of the mission.

"Another amazing milestone is complete for our very soon-to-be project, which is launching American astronauts on American rockets from American soil for the first time since the retirement of the space shuttles," Jim Bridenstine the NASA administrator stated.

SpaceX founder Elon Musk added "It was a picture perfect mission."

What wasn't said was that this test demonstrated without a doubt SpaceX's superiority, at this time, over its commercial competitors. With the inboard flight abort system proven, it verifies the capsule's ability to safely eject the crew away from a failing rocket.

Next up for the Crew Dragon is the launch of astronauts Douglas G. Hurley and Robert L. Behnken, to the International Space Station. That mission is projected to take place sometime between April and June of this year (2020). The capsule itself will arrive at Cape Canaveral sometime in February to be mated with its Falcon 9 rocket. That successful launch will result in NASA's milestone certification of the SpaceX spacecraft for manned space operations.

Instead of conducting a test similar to the one SpaceX has done, Boeing will rely on its previous test data to demonstrate its escape procedures are ready.

NASA's Kathy Lueders, Commercial Crew Program Manager, stated the abort test procedure was crucial to the advancement of manned space flight for both SpaceX and NASA.

NASA's Commercial Astronaut Program Director, Kathy Lueders. Photo courtesy NASA

He likes Rockets

It became obvious after the unquestioned success of SpaceX's escape test flight, that Elon Musk gained a very important fan, President Donald Trump.

"He likes rockets, the President said, you have to give him credit, he's one of our great geniuses, and we have to protect our genius.

He's going to be building a very big [Tesla] plant in the United States. He has to, because we help him, so he has to help us."

The President praised the SpaceX and Tesla CEO at an interview regarding Tesla, the electric car maker on January 22. President Trump also found the space company's regularly successful launches of their reusable rockets impressive.

The president went on to add "He likes the rockets, and he does good at rockets too…I never saw where the engines come down with no wings, no anything and they're landing…I said I've never seen that before.

He's one of our very smart people and we want to cherish those people… Trump added. "He's done a very good job."

Possible Manned Mission in May 2020

SpaceX has checked all the boxes and is ready to launch astronauts perhaps as early as the first week of May aboard their Crew Dragon spaceship. It appears that progress is proceeding so well the space company could likely be ready to launch three months earlier than originally projected.

The Crew Dragon ship was previously launched in March of 2019 and successfully traveled to and docked with the International Space Station then undocked and returned to Earth in a controlled descent. And that is what this mission will be as well, except it will carry a crew of astronauts. With this launch SpaceX will be making space history and put the space company in the forefront of the U.S. effort to resume manned space exploration.

Two experienced crew members will be aboard the launch, astronaut Doug Hurley and Bob Behnken. The two men will be making their third flight into space. This launch of the first commercial crew will mark a milestone for both the space company and the U.S. space program.

Inside the Crew Dragon, NASA photo.

On February 14 SpaceX moved the Crew Dragon to Florida and began prepping the spacecraft for the manned flight. Assuming that all plans come together, the Crew Dragon carrying the two astronauts will launch from Cape Canaveral aboard SpaceX's iconic Falcon 9 rocket.

Astronauts Hurley and Behnken will blast off for the ISS and a rendezvous with space history. A successful flight is a vital element in demonstrating the ship's space worthiness for carrying its precious human cargo into the challenges of outer space. The expectation is that the rocket and Crew Dragon capsule will be an unqualified success and America will be much closer to attaining the goal of returning to the Moon in 2024.

NASA Seeks Student's Help in M2M XHAB

NASA is calling upon our university students for space technology studies and projects centered on space exploration and habitation. The space agency has renewed its M2M XHAB challenge (Moon to Mars Exploration Systems and Habitation Academic Innovation Challenge).

The project rewards work done on space vehicles, robotic missions, human spaceflight, and habitats for the Moon and Mars. The rewards come in the form of money grants ranging from 15 to 50 thousand dollars and include rewards for problem solving research. Although proposals are due by April of

2020 for this round of ideas, there may be more in the future.

Research done through the M2M XHAB challenge could free up NASA scientists to concentrate on other areas of the project.

Chapter 4
A Short History of the U.S. Space Corps

I have no idea what the official insignia of the Space Corps/Space Force will be, but I think it may end up something similar to what you see above.

President Trump envisioned a Space Force, to be the sixth military branch. However, for now at least it's designated the Space Corps and comes under the direction of the U.S. Air Force which makes sense to me, at least for now.

President Trump first proposed the idea of a Space Force in March 2018. The president signed the executive order directing the creation of the new branch later that June.

In the context of this narrative I see the first vital mission of the Space Force as the protection and defense of our Moon. I think this may be the very idea that President Trump had when he put the idea out front.

In the words of the Air Force the actual Space Force (rather than the current Space Corps) has been initially sketched out. At the time of its establishment the cadre[47] of the Space Force will be comprised of 151 Air Force airmen, 24 soldiers from the Army, 14 sailors of the Navy, and nine additional personnel, possibly civilians from the Pentagon and the U.S. intelligence community.

This cadre will form the nucleus of the Space Force and their mission is to organize, create, and build the sixth military branch of service of the United States of America. It is estimated that at least fifteen thousand Air Force personnel and an unknown amount of additional men and women of the other military branches will be assigned.

Understandably the bulk of those assigned will likely come from the Air Force's aerospace

---

[47] A small group of people specially trained for a particular purpose or profession, in this case, military organization.

personnel because of their knowledge of aerospace vehicles. However, soldiers to guard bases and enforce U.S. regulations will be necessary. The Navy's expertise will come into play because I have a feeling that although dubbed the *Space Force*, what we are creating will in fact and in operation, be a Space Navy.

Coastguard sailors will also be necessary, especially at the launching and receiving areas because of their experience in controlling customs and the shipment of goods and personnel.

Of course, where there is a navy, you must also have marines, those ship going soldiers of old. I see that eventually all of the military branches will be represented in the U.S. Space Force.

The Space Force will not only be called upon to thwart threats to U.S. Space assets, but will actually be used to control American's space assets on the Moon, later on Mars, and perhaps even the asteroid belt.

I see the mission of the Space Force to be centered on our expansion into outer space. I see their first mission priority to be the security of the Moon, its space neighborhood, surface and the approaches to it. Later, the Air Force will transfer the duties of maintaining and fighting satellite weapons and anti-weapons and other space based weapons. But I see the Air Force keeping control of surveillance and communication satellites. This mix will no doubt be reshuffled as the Space Force and the Air Force settle in on the assignment of missions.

There will no doubt be jealousy and inter-service rivalries primarily between the Air Force and the Space Force. That is why I see the Space Force, a ship going service, adopting the customs and traditions of the U.S. Navy. And alas, I see my beloved Army giving way to the customs and traditions of the Marine Corps. Thus we will have space sailors and space marines.[48]

If this is the way it goes, I think the rank structure will also be very similar to the U.S. Navy and hopefully for the marines, the U.S. Army. I don't care for the three chevrons and *four* rockers worn by U.S. Marine senior NCOs. It just doesn't seem right! The tradition of the senior service[49] is three stripes and three rockers.

My biggest question is will the congress, that sometimes not so very effective or wise institution, recognize the need for a Space Force and fund it?

But the biggest impact of the Space Force on the Moon will mean we will never lose a war on Earth. The USSF will dominate the very space, the high ground, around our planet.

With Moon bases established at both poles and the Space Force in place, if we have not already

---

[48] If you are a Sci-Fi fan like I am you know that in military science fiction books and films it's almost always a space navy with space marines.

[49] The U.S. Army.

done so it will be time to *claim the Moon for the United States*! There will be no power on Earth that can stop us. But if we hesitate at this most important watershed moment, we may lose our advantage to some other power like Communist China or the Russian Federation.

From the Moon we can grant property claims to the asteroids for anyone who explores them regardless of nationality. Such claims can be enforceable throughout the solar system by the U.S. Customs and Border Patrol, backed up by the USSF, utilizing punitive tariffs or impoundment of offending spacecraft whether robot or manned, against any claim jumping nation.

Establishment of the U.S. Space Force

On December 11, 2018 a bipartisan defense budget agreement was made to create and fund the sixth military branch of the United States, the Space Force. This put an end to the short lived Space Corps. The actual creation of the Space Force as the sixth branch of our military was signed by President Trump on December 20, 2019. The first budget for the newest member of the armed forces was forty million dollars. Two hundred former Air Force members were assigned to the USSF at the same time. And Vice President Mike Pence swore in General Jay Raymond as the new chief of space operations for the Space Force. Over the next year or so officers and enlisted personnel from the other branches, whose jobs are primarily focused on space, will be sworn in as Space Force officers, NCOs, and space troopers.[50]

This is a very important step because it establishes that space defense will come under this new armed force and consolidate it as an independent branch complete with its own Chief of Space Operations. The Space Force is the first new military branch of service since the creation of the U.S. Air Force in 1947.

It is also very important when we lay claim to the Moon, because we will have to be able to defend our claim from any and all challengers. The USSF will especially have to be vigilant regarding the technology stealing of the Communist Chinese.

It is more than probable that most of the space assets of the Air Force and the other military branches will be reassigned to the new USSF. Some examples are the rocket launch site as well as its supporting facilities, operation systems, and personnel at Vandenberg AFB in California. It will likely include the space ground control base at Shriver AFB in Colorado with its facilities and personnel.

Personnel from any and all military branches, many communication and GPS satellites, and the X-37 space shuttle will likely be transferred as well. Eventually it is thought that as many as fifteen thousand Air Force personnel will join the USSF.

---

[50] I started to write space soldiers, but my wife Carol, a long time Army wife, suggested space troopers might be more appropriate, so there you go.

The new branch may also inherit the Near Earth Orbit (NEO) space monitoring mission as well.

The chief of the USSF will report to the Secretary of the Air Force. I believe the increasing presence of our NASA astronauts and their spacecraft and facilities, supported and protected by the U.S. Space Force, will be a force for peace in outer space wherever we establish our authority.

The creation of the United States Space Force may prove to be the most significant thing that President Trump does during his tenure as president as far as the future of our nation is concerned.

President Trump stated "There are grave threats to our national security, American superiority in space is absolutely vital. The Space Force will help us deter aggression and control the ultimate high ground."

Secretary of Defense Mark Esper added "We are at the dawn of a new era for our nation's Armed Forces. The establishment of the U.S. Space Force is an historic event and a strategic imperative for our nation. Space has become so important to our way of life, our economy and our national security that we must be prepared as a nation to protect it from hostile actions.

"Our military services have created the world's best space capabilities. Now is the time for the U.S. Space Force to lead our nation in preparing for emerging threats in an evolving space environment. This new [military] service will… deter aggression… and outpace potential adversaries."

Secretary of the Air Force Barbara M. Barrett has made creating the Space Force her highest priority since she became the Air Force Secretary last October.

"Now is the time for a separate service totally focused on organizing, training, and equipping for space…space today has become far more congested and crowded with other nations and commercial interests actively operating in space.

At the same time space has become a crucial factor in… supporting activities of everyday life ranging from cellphone service, GPS, banking, and the ability to easily and instantly transmit data anywhere in the world. " She said.

On Friday, January 24, 2020 President Donald J. Trump presented the new Space Force seal and said "After consultation with our Great Military Leaders, designers, and others, I am pleased to present the new logo [sic] for the United States Space Force, the Sixth Branch of our Magnificent Military!"

Space Command

Concurrent with the establishment of the Space Force, the Air Force's Space Command was transferred to the new military branch.

The Space Command became the operational headquarters of the U.S. Space Force on December 20, 2019 and became the sixth military branch of the U.S. Armed Forces.

The Space Command was formed nearly four decades ago in 1982 to detect enemy missile attacks on the U.S. and our allies. Later, in 1985, it took over Air Force space launch operations to include the Boeing X-37 space shuttle. It wasn't until 2016 the Space Command developed a Space Mission Force concept to respond to expected future attacks in outer space.

The command is located on Peterson Air Force Base in Colorado. At the time of its transfer Space Command had more than nineteen thousand personnel assigned to include military and civilian. How many of them will remain after the transfer is to be established.

The mission of the Space Command was and will likely remain as the center for National Space Launch; consisting of the launch of satellites and other spacecraft into space using a variety of launch vehicles and operating those spacecraft in space. It will provide space control to ensure the secure use of outer space through the conduct of counter-threat space operations such as surveillance and protection. Force enhancement by providing satellite weather, communications, intelligence, navigation, and missile attack warning.

Space launch operations are at Vandenberg Air Force Base in California and the Cape Canaveral Air Force Station in Florida. Defense program satellites monitor the launch of ballistic missiles worldwide. Its radar systems provide information on the location of satellites, space debris and NEOs (Near Earth Objects).

The focus of this important command center is on recognizing and developing ways that insure our nation's continued unrestricted access to space and to prevent any and all known and potential adversaries from disrupting that.

45th Space Wing

In addition to Space Command, the 45th Space Wing was also transferred to the Space Force. The wing's commander also commands Patrick Air Force Base, Florida and the Cape Canaveral Air Force Station. The wing's mission is to provide access to outer space and to support global and space operations for the Space Force. Depicted above is the 45th Space Wing's unit insignia.

The lineage of the Wing dates back to 1949 showing it was organized just two years after the creation of the U.S. Air Force. At its creation it was designated the Air Force Division, Joint Long Range Proving Ground. The unit has always been stationed at Patrick Air Force Base. During its service the 45th Space Wing as a unit, has earned the U.S. Navy

Meritorious Service Commendation, three awards of the U.S. Air Force Organizational Excellence Award, and twelve awards of the U.S. Air Force Outstanding Unit Award.

Some of the 45th Space Wing's notable launches include: Missions to Mercury and Mars, the Kepler Spacecraft (exoplanet discoveries), NASA Space Shuttle Missions, the Global Observatory, and many others. The 45th Space Wing consists of more than 9,500 military and civilian personnel.

There is no doubt that additional units and personnel will be assigned to the U.S. Space Force in the future.

Military Astronauts

The word is that for the foreseeable future, Space Force Troopers will be earthbound. That is, in my opinion, the wrong answer. To claim the Moon and enforce that claim we need military astronauts. What would be the attributes of a military astronaut?

First of course they must be trained as all astronauts are trained, and then go further. They need to be specifically trained in the use of military weapons in a space setting. What kinds of weapons are we talking about?

The generic term 'space weapon' could be defined as any weapon that can be effectively used in space combat. They could range from weapons used to disable satellites or other types of space craft and would include missiles, laser weapons, and perhaps rail gun cannons or even a space adapted automatic bullet throwing machineguns or small arms. For example the projectiles are launched like mini-rockets rather than fired like standard bullet ammunition.

Strategic weapons capable of attacking targets on Earth or upon the surface of the Moon are likely already available or in the process of development at this time. Defensive weapons capable of eliminating or disabling missiles in flight and reflective shields to thwart laser beam weapons are needed. Missile units will no doubt need to be on the surface of the Moon and in orbit. Placing weapons systems on the far side of the Moon and out of view of the Earth is vital to an effective Moon defense. We would have to establish a 'no fly zone' over the entire far side.

For weapons ideas I suggest we turn to military science fiction and even science fiction based video games for inspiration.

There are hundreds of thousands of young men and women who have developed the quick reflexes and experience of operating space type weapons in the arena of video games and operating drones. These individuals would make good recruits for the Space Force. We will need the best and brightest people, the ones with the 'right stuff'.

The truth is however, that ground combat on the surface of the Moon is a real possibility. There will be enemies from the ranks of such fiction sounding terms as space pirates that prey upon robotic or manned space craft carrying supplies, ships hauling ore from space mining operations, or the kidnapping of space craft operators and passengers. Smugglers will always be with us as well as claim jumpers, enemy agents, and even soldiers. These are very real threat possibilities of the future.

That means that military astronauts will have to understand the challenges of Moon surface combat in vehicles, spacecraft, and even on foot. And they need to be starting that training now, while we have time, before setting up our first Moon bases and colonies. We need to think outside the box, and do it quickly!

During the wars of the 20th and into the 21st century we know it takes ten support troops for every trooper on the ground. How many more may be needed to support our people in space? That's going to call for a hell of a lot of people and we need to start planning for that right now.

Quite serendipitously, KBR Incorporated has been tagged to train a new class of astronauts beginning very soon. What an excellent opportunity, by stretching NASA training funds a bit, to add a dozen or so military astronauts destined for service in the USSF. There is no time like the present, read on.

KBR Astronaut Training

KBR is *the* astronaut training company. This highly successful outfit just inked a Space Act

---

[51] This is KBR's registered trademark.

Agreement with NASA to train astronauts at the space agency's own training sites. This is big news as KBR is a leading provider to the space industry and the first company to offer training for human space operations using NASA as its training field.

I can tell you as a former military trainer for a USAR infantry brigade this is an important development in the training of humans including astronauts. Candidates from the U.S. Space Force would fit right in, and as I mentioned, it would be cost effective and even better training to have them learn beside their civilian counterparts.

In my view an immediate class of Space Force astronaut candidates should take priority or at least have equal training priority over more commercial trainees. The security and defense of our space assets should be prominent in our training thinking. And that goes double when we lay claim to the Moon.

KBR has been a pioneer in space travel for half a century and is well respected in the field. At this time this extraordinary company is operating in nearly a dozen NASA facilities and space centers.

The U.S. Space Force puts its own nametape on a standard U.S. Army issue woodland camouflage uniform. Photo courtesy U.S. Space Force. This Space Force trooper is wearing the shoulder patch for the Space Command, his assigned unit in the Space Force.

The recently created U.S. Space Force unveiled its uniform for the first time on January 17, 2020. Immediately the public wanted to know why it wasn't 'more spacey'.

The message was that the new service's name tape and U.S. Space Command shoulder patch has 'touched down'. The uniform will likely by worn by the many thousands of our military servicemen and women transferred to the USSF.

Despite all the mocking, the uniform is quite appropriate. Space Force troopers will be on the ground for the immediate future. I expect the uniform to adapt as they actually go into space. I find the name tape to be quite attractive and indicative of the future.

The above uniform shows a four-star rank, meaning that this uniform is that of General John Raymond, the first commanding officer of the U.S. Space Force and as of now, the only member of the USSF. Above the service nametape is the Command Space Operations badge.

On the left sleeve of the uniform is the United States Space Command shoulder patch. And above that patch is a full-color American flag patch which is usually in subdued colors (black and green and/or tan).

The Space Force does not yet have a rank structure, a full array of various duty uniforms, or an official name for members of the new service other than my tabbing them 'troopers'.

No doubt General Raymond is enjoying deciding all these things. It's not every day someone gets to determine every detail of a new military service

Things that still need to be decided besides uniforms is a song for the new service to go along with the Army's 'The Army Goes Rolling Along' and the Navy's 'Anchors Away', as well as those of the Air Force, Marine Corps, and Coast Guard. They will also need to come up with a mascot like the other services have.

I'm sure there will be many updates coming along as the new Space Force's traditions are established.

Congress Questions Space Force Commander

Members of the House of Representatives on the Armed Services spoke to General John Raymond on March 4, 2020 regarding Space Force operations such as reserve forces, space launches, and future tasks to organize the Space Force itself.

They got few detailed answers as the general stated he was working to 'round out' the new military branch its capabilities and its manning levels. Most of the general's answers were vague because what the Space Force is doing is mostly classified.

In my opinion this is because the mission of the Space Force is still evolving. I would think that there are at least two camps with varying views. I suspect that the Air Force would like to see the subordinate Space Force continue in the same manner they were while still part of the USAF. But there is likely pressure from national command authority, through General Raymond, for something much more.

There is almost certainty a desire for the new space oriented military to get into space. I have mentioned in a different part of this book (see KBR Astronaut Training) that I believe we need to train strictly military astronauts, Space Force troopers (my designation) who are learning the trade of the military which includes of course, space warfare.

The general gave a hint in that direction when he stated the branch's budget is for space fighting domains "…a strong pivot to space as a priority…"

Certainly General Raymond would not want to speak of such matters to members of congress outside of a classified setting.

On the matter of new personnel joining the USSF, sixty-five new graduates of the Air Force Academy are slated for assignment to General Raymond's command.

The creation of the Space Force has engendered a new level of excitement, as indeed has all the new emphasis placed upon all things space oriented in our young people. It is a good time to harness that excitement, to encourage an interest in the hard sciences and to bring forth the military leaders, the scientists, and the engineers we are going to need for the future.

Without a doubt many Americans envision members of the Space Force as Star Wars like storm troopers or Heinlein's star ship troopers and Star Trek's Federation of Planets. Refer to those oh-so-American ideas and encourage them. They can light the fire of a determined national service for our youth, and we are in need of that.

Space Development Agency to Join Space Force

Derek Tournear, the director of the newly formed U.S. government's Space Development Agency, stated the SDA will join the Space Force in October of 2022.

Formed in March of 2019, the Space Development Agency's mission is to create and procure satellites needed to accomplish our national space operations both military and civilian in nature. Waiting until 2022 to join the Space Force allows the SDA time to obtain and have satellites actually operational before the merger. The agency is expected to have satellites launching by September of 2022 and be operational for Earth based units to use.

In order to reach the 2022 deadline, the Space Development Agency will move quickly to solicit and award satellite contracts to space companies able to respond almost immediately with spacecraft construction.

The agency welcomes contract bids from satellite providers that can meet the stated deadline in eighteen to twenty-four months.

Space Force - Space Academy

During President Donald Trump's recent 'State of the Union' address General Charles McGee, (100) a former Tuskegee airman of WWII fame, was introduced. In addition his young thirteen year old great grandson, Iain Lanphier, was also introduced. The young man has stated his desire to follow in his famous great grandfather's example of military service to our nation; he wants to join the Space Force.

Although there has been no talk of establishing a Space Academy separate from the Air Force Academy that could become a necessity sometime soon in the future. I have spoken to a few young people and the idea of joining the Space Force is exciting to them.

Should such an academy come to be, I'm sure Iain will be one of the first to apply and it would seem fitting that any such academy be on the Moon.

China's Reaction to the Space Force

Almost immediately after the official announcement our sworn adversary, Communist China, communicated its displeasure in the creation of the U.S. Space Force. That alone is a good indication that we are moving in the right direction.

Foreign minister Geng Shuang characterized the US Space Force as a "...direct threat to outer space peace and security." and then added the Communist state is "...deeply concerned about it and resolutely opposed to it."

Of course they are, because the Communists are rushing to get their Moon and Mars occupation program quickly off the mark.

China contends that "...U.S. actions are a serious violation of the international consensus on the peaceful use of outer space, undermines global strategic balance and stability, and pose a direct threat...."

If the communist regime is this concerned, then of course we are doing the right thing at the right time. Let there be no mistake, space has been and will continue to be a war fighting domain. And as such, we must be ready to conduct military operations up to an including combat in that sphere.

Only a few days after voicing their 'concerns' Communist China launched a new space mission to test the capability of its 'Long March 5' rocket.

The stated purpose of the rocket's launch is to test its ability to haul a rover to Mars, send a probe to the Moon, and deliver modules to construct a planned space station. Of course other tasks of a military nature are also quite possible and quite likely.

2016 file photo of the Long March 5 rocket. Credit AFP.

The Long March 5 stands nearly 19 stories tall, has four strap-on rocket boosters and develops 10,631 Kilonewtons of thrust. The launcher is rated to carry up to 25 tons of cargo and personnel. The LM5 rocket is essential to China's very ambitious plans for a mission to the Red Planet next year and their hope of creating a crewed space station by 2022.

The last attempted launch of a LM5 in July of 2017 failed. The failure impacted their Moon program and put them behind by eighteen months. The LM5 accounts for only a small part of the billions of dollars invested in the Communist Chinese space program in its attempt to gain parity with the United States. China spends more than Russia and Japan on its civil and military space programs and is the third nation to put a human into orbit.

In a rather impressive accomplishment, China landed a rover on the far side of the Moon dubbed the Chang'e-4 lander. I can assure you it was no coincidence the rover was set loose on the Moon's South Pole-Aitken Basin near where water was found by U.S. space efforts. In January of 2020 it noted its first anniversary on the Moon.

China will also seek to build an international lunar base, possibly using 3D printing technology, another American innovation.

It should be clear to anyone reading these words that Communist China is not only our adversary on Earth, but also in outer space. Quite indicative of this is the fact that China and the Russian Federation both have their own space force, although neither of them call it that.

The Chinese Communist's People's Liberation Army has had a space force for the last four years. It's actually called the PLA Strategic Support Force. Assigned areas they are responsible for includes outer space, cyber space, electronic warfare, and psychological warfare.

China's president Xi Jinping has every intention of seeking a land grab in space and of course the closest land to grab is our Moon. This intent is reflected in the following statement from the Chinese state council made in 2015.

"Outer space has become the new commanding heights in international strategic competition... Countries concerned are developing their space forces and instruments, and the first signs of weaponization of outer space have appeared."

Competition in space from both China and Russia is without a doubt directed at the United States because we have in the past and are currently leading the space race and have a legitimate claim to sovereignty over the Moon.

Of our two large world adversaries, China's recent space accomplishments have outstripped the Russians and they have moved to the forefront in challenging our space dominance. That only reiterates why we must renew our efforts and quickly.

The recent launch of the LM5 rocket has put the Chinese space program back on track toward Mars and its intended collection of soil samples from the Moon. Additionally, their location system, similar to our GPS system is nearing completion, and that will give them a better grasp and understanding of the physical Earth.

In a thinly veiled propaganda film titled *The Wandering Earth*, the Chinese released a science fiction movie in which Chinese space program astronauts save the world from imminent destruction.

Outer space remains peaceful for now, but just a few years ago China launched a missile that actually destroyed one of their own inoperable satellites.

The Chinese space force or PLASSF is refining their military space operations and has conducted related research for the past four years. They rival both our Space Force and its operational U.S. Space Command.

These are the same Chinese Communists Democrat Joe Biden, presidential candidate, described in May of 2019 in this manner: *"They're not bad folks, folks. They're not competition for us."*

The well respected Asian security analyst, Gordon Chang, said this about former Vice President Joe Biden's campaign statement: *"He's absolutely incorrect,"* Chang declared, adding *"...you know, one can argue that China doesn't pose for instance a robust challenge to the US, we can argue that we will prevail. But to say as the vice president said that they're no competition to the US is just inexplicable."*

If elected, this is the man who will hold sway over our space program at its most critical of times. We need someone who is aware of what's actually going on in the world, someone who has shown the necessary intelligence and chutzpah we need.

In this real life space opera not only are the Communist Chinese not good folks, folks, they are the bad guys.

The Cost of a USSF Moon Base

It's not easy or cheap to liftoff from the surface of the Earth and fling our eager spacecraft into the vaults of heaven and on into outer space. In fact, according to NASA accountants, it costs ten thousand dollars for each pound of weight launched. So ten pounds of dried beef or other protein would cost a whopping one hundred thousand dollars! The good news is that NASA is working on cost cutting innovations that will, eventually, bring the cost of rocket launches down.

As with humans living anywhere, life on the Moon will require vast amounts of energy. There are areas on the Moon where the sun shines continuously and are ideal for solar panels. Solar panels are one way to get power but they will have to be augmented by other sources. The current idea, as outlined earlier in this narrative is the Kilopower project's nuclear reactor. The estimated cost of this project is about twenty million dollars.

Hauling food to the Moon initially is a must. But while we are doing that we need to be able to eventually produce food on the Moon. Enter hydroponic farms that will trigger another need, vast amounts of water. Luckily there are vast amounts of water on the Moon. Transporting it to where it's needed may prove to be the big problem we have to solve. Moon gravity is about one-sixth of the Earth's, so water will flow slower. Powerful water pumps will be vital to water production.

Lava tubes on the Moon could provide all the space needed for hydroponic farming assuming they can be closed up air tight. Growing lights would be required because there would be no direct sunlight. But direct sunlight without an atmosphere filtering it would destroy plants anyway. So the lava tube route is likely the best way to go. We will know more on the feasibility of this approach once we begin exploring the caves of the Moon.

Lava tube use may also be the better way to shelter people from the threat of radiation. Without a doubt however, some facilities will need to be on the surface. The cost of such facilities will more than likely be expensive although using the technology of huge 3D printers may alleviate that some. The printers could use materials readily available on the Moon such as the crust known as regolith. Regolith is the fine dust of volcanoes, asteroids, and meteorites that have stuck and pounded the rocky crust of the Moon. This should make fine cement when blended with necessary strength enhancing additives.

As mentioned earlier; water that is already on the Moon may be refined to produce hydrogen for fuel, and oxygen for breathing.

Without a doubt colonizing the Moon will cost a great deal of money, and yet the rewards of possessing the Moon as a U.S. territory will provide future wealth literally in the trillions and likely in quadrillions of dollars.

The surface area of the Moon is equivalent to gaining another continental land mass approaching the size of Africa! This kind of expansion has not happened since the discovery of the western hemisphere of our world. We must not miss this opportunity by being timid!

One option the USSF may have to use when building a Moon base is the Bigelow Company's Aerospace Inflatable Habitats.

The Bigelow BA-330 Inflatable Space Habitat

The 330 is a reference to the interior space of the habitat which is 330 cubic meters in size which equates to 11,650 cubic feet about one-third of the ISS. It can support up to five people for several months and is designed to be a self-sufficient space station. A smaller Bigelow unit, the Expandable Activity Module, was attached to the ISS in 2016 for use and testing.

Bigelow's hope is that NASA will select the BA-330 for use on the Moon orbiting space station, known as the Lunar Gateway. Construction of the new space station is expected to begin in 2022 and is part of the planned Artemis Moon program.

Eventually Bigelow envisions its inflatable habitats becoming integral parts of any Moon bases and will also be included on the trip to Mars and beyond.

To claim the Moon we need and we must have military astronauts and military spacecraft to protect and defend the U.S. Moon!

Space Weapons

Let's talk a little bit about space weapons. On June 25, 1974 The Soviet space station Salyut 3 was launched. This space station, manned by cosmonauts was armed with a 23 millimeter caliber cannon. While operating in space the cannon was test fired, successfully, at targets ranging from 500 meters and up to 3,000 meters away. It goes without saying what this quite conventional weapon, used in the non-conventional setting of space, could do to satellites, spacecraft, and other space stations like the ISS. So, from very early on weapons were deployed to outer space and their use demonstrated.

This cannon is far away from the weapons being dreamed about, on the drawing board, and likely, already created and being tested. More exotic space weapons are hinted about or even the subject of rumor and much speculation, no doubt much of it well founded.

One weapon that has been talked about was thought of by science fiction writers Jerry Pournelle and Larry Niven in their book *Lucifer's Hammer*. The concept was that metal telephone pole sized rods could be dropped from outer space with the impact on Earth of a nuclear weapon. Such a weapon would go straight down on its target and impact without defensive radar ever seeing it or seeing it so fleetingly that no defense could be triggered. Supposedly the U.S. developed such a weapon in a weapons program called Project Thor. However, that remains unverified.

These and many other space weapons will likely come in a future that speaks to the fact that we need a Space Force, in space, to guard us from them.

CHAPTER 5
Regulating Outer Space

'You need regulation of space — there isn't 'a cop up there'

Adrian Steckel CEO of OneWeb

Death in Space and on the Moon

To this point we have spoken of life on the Moon, but everyone is aware of the opposite of life which is, of course, death. How will we handle the demise of humans both in space and on the Moon?

NASA has thought about this, because inevitably, it will happen. There are no bacteria on the Moon to decompose bodies buried on its surface. Bodies would actually become freeze dried mummies. Here is the proposed procedure should the family or the deceased desire the remains be returned to Earth.

The body is placed in an airtight body bag and stored in an airlock. With no heating in the airlock the body would be subjected to the 'normal' temperature of space of -270 degrees Celsius and would both be dehydrated and frozen solid. So far so good, but the next steps seem a little disturbing, or not, depending upon your personal view.

The body would then be shaken by some means of vibration, mechanical in nature, which would result in shattering the remains to pieces producing a somewhat heavy dried powder. This would eventually be returned to Earth and the deceased's family.

Personally I think the deceased could be simply frozen and returned in tact to our world.

Another option is similar to burial at sea if the body is not returned to Earth. This could be as simple as putting the body in a bag and letting it drift off in space, perhaps eventually to be tugged by gravity back to Earth and cremation as it passes through the atmosphere.

However, should the person die while on the Moon, then more 'normal' burial practices may be used. That means there is a need for people with mortuary affairs skills assigned to Moon bases and colonies.

Moon Invasion and Contamination

By definition any living thing that arrives on the Moon is an invasive species. That means that contamination of the lunar environment is assured. This also means we will need to safeguard against any uncontrolled release of biological, radio-logical or chemical contaminants.

A case in point is the failure of Israel's 2019 *Beresheet*[52] lunar lander mission that resulted in the crash of the lander. Aboard the lander were thousands of microscopic tardigrades, more popularly known as *water bears*, a very robust creature that exists in its trillions on Earth. These creatures are surprisingly hardy and live in just about any environment including extreme ones. When their metabolism slows down because of adverse conditions, they can hibernate for long periods of time.

After the crash NASA made a somewhat understated statement that reads "Uncontrolled biological contamination of the Moon's surface is not scientifically ideal...".

It would seem the results of this contamination will likely prove of no importance. But it does point out, that accidents can indeed happen.

The other side of this coin is the possible disturbance of slumbering pathogens, creatures, or artifacts already present on the lunar surface. The Moon is as ancient as the Earth and subjected to the same bombardment of natural missiles. Ergo, if life generating organisms land on Earth, they also landed on the Moon and may simply be dormant, waiting for something, or someone, to wake them up. This all sounds rather ominous doesn't it?

Enter the theory of Panspermia that states that '...*life on the earth originated from microorganisms or chemical precursors of life present in outer space*

---

[52] Beginning in Hebrew.

and able to initiate life upon reaching a suitable environment.'

We don't think the Moon provides a suitable environment for the starting of life. However, humans through their actions of lunar colonization could unwittingly supply such an environment.

One needs only to watch the 2011 science fiction horror movie titled *Apollo 18* to see something along these lines.

NASA did in fact schedule an eighteenth Apollo mission but it was scrubbed in the 1970s along with Apollo 19 and 20 due to budgetary considerations. In the movie however this eighteenth mission was launched in secret at the request of the Department of Defense with a crew of two astronauts.

Rock specimens collected on the Moon are subsequently revealed to contain a deadly creature that, of course, kills the astronauts.

Although fiction it does make a cautionary statement even if it is a farfetched one.

First Orbit Cleanup Spaceship

Since the launch of the Soviet Union's Sputnik satellite in 1957 the world has orbited thousands of satellites around the Earth. More than half of them still remain in orbit today with about half of those fully operational. The rest are dangerous defunct missiles whizzing around our planet at thousands of miles per hour. They are a hazard to space navigation and pose a risk of destroying still working and valuable satellites and spacecraft, including the International Space Station (ISS).

This same scenario is likely to happen around the Moon, and very soon. Space faring nations are scrambling even now to plan manned and unmanned missions to the Moon replete with orbiters, landers, and crash landings. If action is not taken to regulate the use of space around and adjacent to the Moon it will quickly become another space junkyard.

Currently only thirty percent of satellite operators adhere to guidelines that request they retire satellites after twenty-five years of use. New spacefaring nations may not feel an obligation to adhere to U.S. regulations. Currently there is no internationally recognized space traffic control that regulates Earth orbiting satellites much less satellites orbiting the Moon.

So, who is to regulate the usage of this soon-to-be vital space? Why us, the U.S., of course. Would we tolerate any other nation doing it? Of course not, we were the first on the Moon and we need to claim and take charge of it and regulate it and its space byways.

Specifically it is the Space Force who needs to take responsibility for both regulation and maintenance of the space byways. For example, depleted satellites that cannot be recovered could be used for target practice by Space Force warships. Or, they could be rigged for detonation or removed by a rocket assisted shove into deep space. Of course this cannot simply be done willy-nilly or these objects floating around in the solar system could come back to haunt future space missions as hazards to navigation.

This would require at least two moon bases complete with radars, interceptor spacecraft, space faring tugboats, missile launchers, tracking computers, barracks, and many other vital and necessary systems.

Other federal agencies will have to be involved including U.S. Customs and Border Patrol and other regulating agencies that controls trade, communications, and energy. Quite frankly, nearly every federal agency will have a hand in this unless it is streamlined.

Japan Enters Orbit Cleanup Effort

Japan's Aerospace Exploration Agency (JAXA) has selected startup company Astroscale for its first orbit debris cleanup mission. Astroscale has focused their company efforts to address the continuing probability of more and more space junk accumulating in low Earth orbit. Its first target is the remains of a large spent Japanese rocket. The company will begin by building and launching an inspection satellite to survey and map the rocket's remains in 2022.

Already involved in other debris removal efforts, Astroscale will demonstrate its evolving removal process sometime this year.

ClearSpace-1 Mission

Thanks to the European Space Agency (ESA) the first spacecraft designed to remove dead satellites and other space debris will launch its initial mission in 2025.

The Swiss company ClearSpace produced the space faring vehicle to locate and capture large pieces of space debris. Once a target is secured CS-1 will tow it into a decaying orbit where it will enter the atmosphere and safely burn up.

Its first target is part of a spent rocket weighing 265 pounds that entered orbit years ago and no longer serves a useful purpose. This kind of removal is termed a 'non cooperative capture' that means it not designed to be captured. As the process is refined it is hoped that orbital equipment will be equipped with 'handles' that allow for easier capture once the satellite is no longer operational. But that will require the voluntary cooperation of spacecraft manufacturers and space launch companies.

ClearSpace is hopeful that active debris removals will soon become a standard part of space operations. Orbiting space debris is an increasing problem that has been building since the launch of Sputnik-1 in 1957. According to ESA records there exists more than 34,000 pieces of inoperable satellites and spent rocket parts such as boosters larger than ten centimeters across still in orbit. This debris is a hazard to navigation and could actually damage or destroy space craft. Even the International Space Station was threatened in the past from space debris.

Space companies and agencies like SpaceX and the Indian Space Research Organization routinely launch multiple small satellites numbering in hundreds on one rocket launch. So the space debris problem is likely to grow.

Looking at possible legal liability for damaging spacecraft, some space contractors are seeking a solution to this problem. ClearSpace is hoping to have the solution already available.

Perhaps an American company will step forward to help clean up the orbital lanes as well.

A Space Scare over Pittsburgh

Near the end of January 2020, space scientists voiced the concern that two no longer operational satellites had one chance in twenty of colliding. One of the satellites is a dead NASA telescope and the other is a now useless U.S. Air Force spacecraft.

The collision was projected to happen on the 29th when the orbiting satellites were both passing 560 miles above the city of Pittsburgh. The concern was that if indeed they did collide, the resulting debris field would pose a threat to future space launches around the world. The collision did not take place, but what would be the result if they had?

It should be noted that the Air Force, who studies Near Earth Object (NEOs) has not predicted these two satellites are in danger of colliding.

Had the satellites collided with one another a large debris field of up to 300,000 hurtling pieces of metal, basically a deadly cloud of bullets could be created. These things would remain in orbit for decades, hazards to other spacecraft and to navigation.

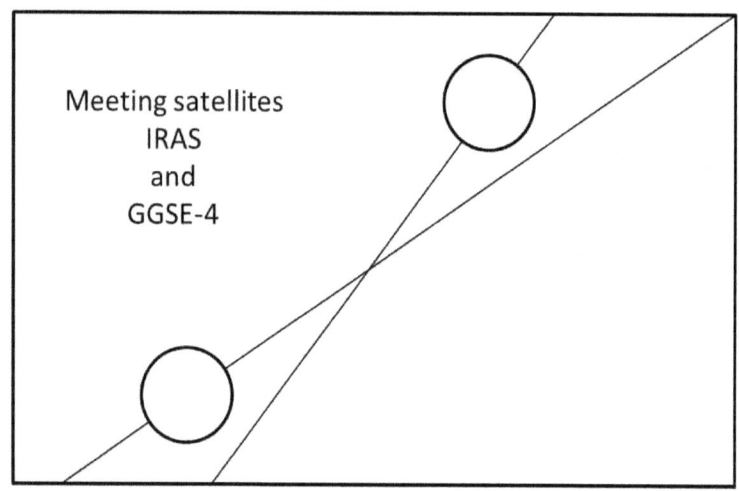

Image by the author.

Over time the various sized pieces of debris could spread out as gravity works its way to disperse them and possibly result in an ever growing menace. This could be the same fate around our Moon as well if space regulation is not initiated.

We were lucky this time because the dead and useless space machines missed one another. But that might not be the case the next time they approach one another.

NASA illustration of the IRAS telescope.

End of Our Manned Spaceflight Drought

In July of 2011 NASA retired the space shuttle, and since that time we have had to rely on the Russian's space program to get to orbit. Private American companies have been developing commercial crew operations and since September of 2014 NASA has invested billions of dollars in the companies of SpaceX and Boeing to finish work on space capsules.

Now, the SpaceX Crew Dragon, Boeing's Starliner, and Blue Origins New Shepard spacecraft are nearing completion and certification for human space flight. Expectations are that early in 2020 we could be ready for launching a manned flight. If that happens NASA astronauts Doug Hurley and Bob Behnken will lead the way to a new era of manned space exploration as they travel to the International Space Station.

Emergency Earth Rescue

In the future there may come a time when Earth may need help to survive a catastrophic event. One such threat of course that has been explored both by Hollywood as well as space scientists is the threat of an unavoidable asteroid strike.

Another is the eruption of a major volcanic event such as that of the Yellowstone area of the U.S. and a myriad of other possible disasters. There are a large number of Earth catastrophes you can explore online. The Moon could be our best answer to nearly all of them.

Once we are established on the Moon and our progress is sufficient we can begin to make the Moon an emergency haven for humanity. A civilization on the Moon could offer rescue to at least enough people to maintain a viable population to reestablish our home world if necessary once the emergency is past.

To do this we have to plan for the future, turn lava tubes under the lunar surface into habitats that offer protection from solar and cosmic radiation. We will need to store food and other necessary items to support a large population for a long period of time. How large and for how long will be dictated by the nature of the emergency on Earth.

Obviously this is a worthwhile and necessary development for human survival.

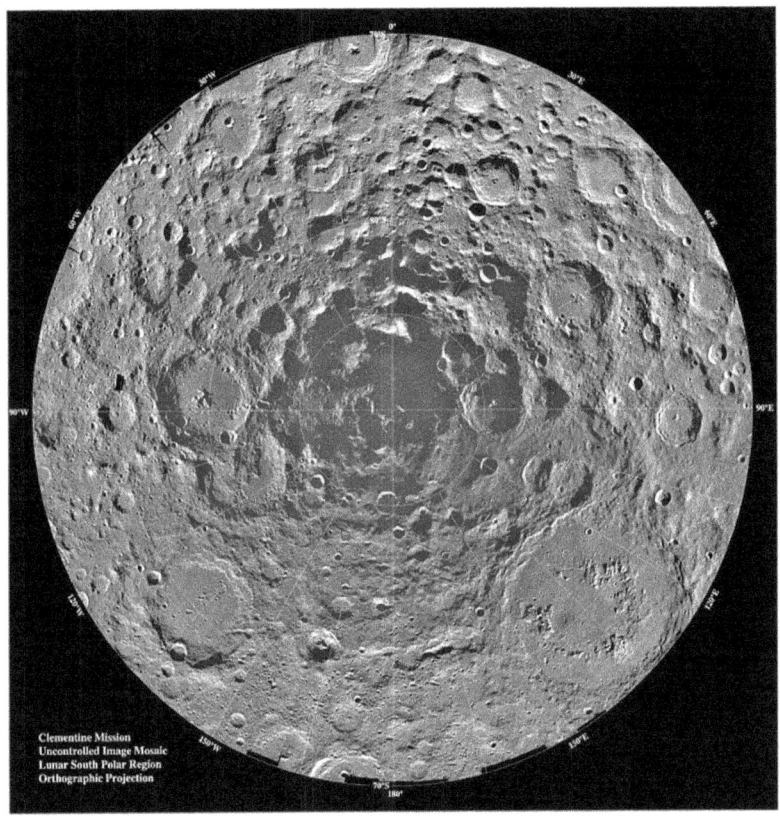

The south pole of the Moon. Here, there is ice. (NASA)

Chapter 6
Return on Our Space Investment

Since its inception the U.S. has spent $601.31 billion on NASA. The U.S. has the largest budget for space exploration by far, spending over six times more than Communist China. We have led all countries in the exploration of space since 1958, the year after the USSR launched sputnik. These facts alone justify our claim to the Moon.

Procure Space, a space investment exchange traded fund for companies involved in the space economy recently made the following statement.

"Mainstream financial markets are only just starting to awaken to the commercial and disruptive opportunities that space offers, as technology is starting to tear down the high entry barriers to access space. We forecast that the combination of declining space launch costs and advances in satellite technology will raise the value of the space economy from $340 billion currently to nearly $1 trillion over the next two decades."

This statement can be considered conservative as Procure Space did not include very important private companies like SpaceX and Blue Origin. These two companies are on the forefront of reusable rocket development, low orbit spacecraft and evolving communications. All of these efforts will most assuredly expand public enthusiasm for pushing ahead in space exploration and investment in its future.

Others have estimated that the space industry could top 40 trillion dollars over those same two future decades starting now. Frankly, I believe that estimate to be woefully low, maybe even ridiculously low. This is the 1849 gold rush, the Klondike gold rush, the opening of the West to settlement, the rise of the clipper and steam ships, and the cutting of the Panama Canal to join two oceans all rolled into one. And it will eclipse all of them put together. If you have money to invest, invest it in space technology and exploration.

The exploration of space will require the mutual support of both government programs and private industry. Success will call for close interaction and cooperation on all levels of human endeavor united together like never before. In short, it will take a World War II kind of effort. We must substitute our courage and willingness to sacrifice in a different kind of all out struggle for mission accomplishment.

Communications of all kinds will need to expand and fill new needs as they develop. There are many communications companies but there will no doubt be a need for even more with new ideas and new technology to speed communications and use them in unthought-of ways and challenging new places. For example, communicating from the far side of the Moon.

This is only one of the many needs coming up like off Earth construction, transportation to, on and around the Moon and other celestial bodies. The process of securing water and all the other materials needed for a growing and thriving space civilization.

The Department of Defense has established the Space Development Agency to facilitate and coordinate the necessary mix of government and private sector space endeavors.

The New Gold Rush

The moon is our gas station in space and will fuel our rockets to the asteroids, Mars, and beyond. The lag in communications between Earth and the Moon is only three seconds, much faster than it will be from Mars and other outlying points. That will allow our robots including rovers and many other types of mechanical and electronic machines to receive near instantaneous instructions. We will be able to erect habitats assembled by robots.

Moon Facilities Available for International Use

There are facilities we can initially make available for lease to our international partners and charge fees for such use and services. Here are a few that are obviously needed shortly after we settle on the Moon.

Spaceport facilities have to be constructed, with designated landing places for manned spacecraft, and terminals for the handling of passengers. Additional terminals with storage facilities will be necessary for cargo missions.

Nations wanting to place satellites in orbit around the Moon will apply to NASA on Earth. The application must state their mission and why they are deemed necessary. Their satellites will have to be inspected prior to launch, possibly during every phase of their construction, by U.S. customs agents or the Space Corps to insure our national security. These services of course will have to be paid for.

Lunar land for construction of bases of any kind will be inspected by US geologists and engineers before being leased. Land will never be sold. All construction will be supervised by U.S. lunar engineers and security people. All bases are subject to periodic unannounced security checks. There will be no secret bases or laboratories on the Moon for any one besides the U.S.

The U.S. will have sole rights to manufacture and lease lunar equipment of all kinds including all rovers and movers. International drivers of such leased vehicles will require training and certification by our own drivers and pilots.

Launching spacecraft from the Moon for other outer space destinations will also come under U.S. control.

Eventually we will need to establish a US Space Academy for training our military, the military of allied nations, and all civilian pilots. This academy will come under the administration of the Space Force.

I also can see the need for the establishment of a human genome project to safeguard DNA should a catastrophe befall Earth. This will truly establish humanity as a two planet species.

All pertinent laboratories for research will be established and administered by U.S. personnel. Observatories will be constructed and come under the same scientific administration as the laboratories.

Any Lunar city colony will come under U.S. laws. This will require a court system and a department of justice with its own attorney general and U.S. attorneys. Humans being fallible there will no doubt be a need for a place of incarceration.

Medical facilities for treatment of humans living upon the Moon and for medical cases referred for treatment to Earth.

A department of lunar agriculture is needed for the establishment of hydroponic and other farms to produce as much food as possible in a lunar environment. An energy department is needed for the production of electrical power and rocket fuel for industry and transportation.

We will no doubt have babies born on the Moon. Careful consideration of the effects of raising children in low gravity will have to be researched and procedures established. It may become necessary to establish high gravity nurseries for a type of 'incubation' for new born babies. This could take the form of a type of gravity producing centrifuges in orbit.

Americans born on the Moon will of course be U.S. citizens, but those of other nations will not be. Laws and regulations establishing this will have to be promulgated. This is only one of many things that will have to be addressed.

Lockheed Martin Space Revenue Growth

The Lockheed Martin Space Division of its business grew over four percent during 2019 due to its work on NASA space projects, military systems, and missiles.

The space division is building the Orion space capsules for NASA to send astronauts to the Moon and perhaps later, to the Red Planet.

This large space company for over forty years has provided the aeroshell used by NASA for its planetary rovers. In December, they completed the aeroshell that will protect the 2020 Mars Rover, scheduled to launch in July, as it lands on the distant planet's surface.

Lunar Natural Resources

The resources available for our use will take very hard and dangerous work to extract, but the potential rewards will be very great indeed. As we look forward to traveling deeper into space the Moon will be the optimal place to launch from to travel to the asteroid belt, Mars and on to the planet sized moons of Saturn and Jupiter.

Helium-3 is a possible energy source more available on the Moon than on Earth. The fusion of helium-3 atoms releases large amounts of energy creating only small amounts of radioactive material. Helium-3 is likely embedded in the top layer of the Moon's regolith borne by the solar wind that has struck the surface over billions of years.

Only the gas giant planets of our solar system are thought to have Helium-3 in larger amounts than the Moon, at least at the time of this writing.

Water frozen at the poles is a precious resource and as outlined earlier it provides oxygen for breathing and hydrogen for rocket fuel.

Rare Earth metals, chemical elements of the Earth's crust, are vital to many technologies.

This includes electronics, computers, advanced transportation, clean energy, healthcare, network communications and national defense. These elements are known to be available in certain regions of the Moon. Most of these elements on Earth are located in China, who has a monopoly on them.

These elements also make technologies weigh less, reduces emissions as well as energy consumption, miniaturization, and numerous other advantages.

Platinum elements are available on the Moon and go for thousands of dollars per kilogram. These elements can be extracted from iron meteorites that have impacted the Moon and number most likely in the hundreds of thousands of tons.

At the beginning of this narrative I asserted that the Earth and Moon are essentially comprised of the same 'stuff'. This suggests that precious metals should be available, no doubt in varying amounts, for mining on Luna. Yes, I mean gold, silver, platinum, titanium and more.

In its past the Moon had a volcanic life, this suggests that diamonds could have formed and may be present today. Perhaps a little lunar spelunking would reveal veins of precious metals and gems waiting to be scooped up.

Lunar basalt found mostly on the near side of the Moon contain up to 20% titanium usually in the form of the mineral ilmenite. Lunar ilmenite is an iron titanite, containing ferrous iron, titanium, and oxygen.

There are undoubtedly many other resources likely to be discovered on the Moon. The Moon appears to be a gold mine in more ways than one. And it will be there forever or at least as close to forever as we will likely ever know it. These treasures are not going away in any foreseeable future.

There is nothing more conducive to the future commercial development of the Moon, than the fact that it is under the regulation and control of the United States.

This artist's-concept illustration depicts the spacecraft of NASA's Psyche mission near the mission's target, the metal asteroid Psyche. (NASA)

The Trillion Dollar Metal World of Psyche

A mysterious and interesting asteroid has caught the attention of NASA to such a degree, that a 2022 mission is scheduled to go there. Designated Psyche 16, scientists believe it is the metal core of a planet that failed to complete its formation. The theory is that its creation was rudely interrupted when its rocky mantle was destroyed in an asteroid collision.

In Greek mythology Psyche, a nymph, was made immortal by Jupiter at the request of her husband Cupid. Psyche was a mortal woman but when Cupid fell in love with her, he took her from Earth.

Jealous, Venus killed Psyche but was thwarted in her terrible deed when Jupiter interceded.

This asteroid is named for the beautiful Psyche and holds out the possibility of great fortune if you can get to her. Its diameter is about 1/16th of the Moon, or 140 miles. It is thought that Psyche is likely made up of metals such as iron and nickel. An early planet perhaps initially the size of Mars, its outer rocky shell may have been knocked off in a series of devastating collisions in its distant pass. Psyche orbits the Sun between Mars and Jupiter in an orbit that ranges between 235 and 309 million miles from our star. The asteroid rotates every four hours but takes five years to travel around the Sun.

The mission to Psyche is scheduled in 2022 with a projected arrival in 2026. The spacecraft will study Psyche's physical make up and map it using a spectral imager, a neutron spectrometer, a radio transmitter and a magnetometer. It is expected that the study will determine if the asteroid is in fact the remains of a planetary core.

What makes it so potentially valuable is the asteroid is projected to contain tons of gold, silver, and platinum. Since its rocky mantle is gone, the harvest of these precious metals may be, relative to other asteroids, a matter of getting there and loading it up on a spaceship designed for mining.

If you are an entrepreneur you might want to start building a mining spaceship right now and steal a march on the competition. Space mining could be the next 'big boom' and lead to unimaginable economic rewards to forward thinkers.

There is a bonus to mining Psyche or any other asteroid or moonlet. While relieving your wandering space ball of its mineral treasures, your remodeling crew is busy shaping the hollowed-out rock into a spaceship.

In space you don't have to be aerodynamic. And, in case your space rock is composed of the 'right stuff' you would leave a thick enough outer shell to insure air tight sealing. Once that is done add a few steering rockets to the outside and a space lock, bolt on a rocket motor for the main propulsion and deliver it to your customer. You know, your customer, the space company who just paid you a half billion dollars or more for a ready-to-go space freighter.

On February 28, 2020 SpaceX was awarded the $117 million dollar launch services contract to go to Psyche in 2022. This mission appears to be a very reasonably priced expedition for potentially very valuable information. A Falcon Heavy rocket is scheduled for use and will launch from Cape Canaveral, Florida.

The trip to Psyche will be a formidable one as its orbit lies between Mars and Jupiter at least 235 million miles away. That is its closest approach to Earth and the Falcon will likely launch on that closest of possible trajectories.

The lead team for the Psyche mission is from Arizona State University in Tempe, Arizona. The principal investigator for the Mission, Lindy Elkins-

The Abduction of Psyche [by Cupid] a painting by the 19th century artist A.W. Bouguereau. Is she waiting for you?

Tanton stated "...we are one big step closer to uncovering the secrets of Psyche...and that means the world to us."

California's Jet Propulsion Laboratory (JPL) will manage operations for the asteroid exploration mission.

This is NASA's first use of the Falcon Heavy for a mission and will without a doubt be an exciting event for space fans. I know I look forward to viewing the film footage and witnessing a priceless addition to our knowledge of the solar system.

Atomic Probe of Moon Dust

ATP (Atom Probe Tomography) was recently used on Moon dust to determine conditions on the lunar surface and in the subsurface as well as for the presence of helium-3, water, and any other elements.

The technique analyzes single grains using a focused beam of charged atoms. The process chips single atoms off the sample and analyzes them one by one. The atomic weight of these atoms identifies what element they are composed of.

Researchers can then create a Nano scale map of the Moon's surface as well as its subsurface. That information shows what areas to mine for valuable minerals and elements. This technique works with samples gathered from other sources in space too like asteroids and perhaps even comets.

Moon Penal Colony

Let's face it, we have some horrendous criminals on Earth, some of them incarcerated and others that are not. We seek to isolate them, get them away from our societies and protect our citizens from them.

All nations have these kinds of problems, what to do with serial killers, leaders of genocide attacks, and war criminals short of execution? Separate them from the very planet of their birth, house them in a penal colony on the Moon. There they could be housed for the rest of their lives and upon death, disposed of, leaving no martyr's marker behind.

Of course it would be expensive, not only housing them, but transporting them to the Moon, and guarding them. The nations involved would have to be ready to separate with a great deal of national treasure. But on the bright side, they would never have to think of these monstrous criminals ever again.

First Satellite to Satellite Refueling

The life of a communication satellite was extended by the first rendezvous and refueling in orbit. The 19 year old Intelsat 901 was visited by a Northrup-Grumman MEV-1 (Mission Extension

Vehicle – 1) more than 22,000 miles above the Earth on February 26, 2020. In an impressive gentle docking the MEV-1 gripped the failing satellite and will remain with it for the next five years providing it energy to continue to operate.

To facilitate the docking, Intelsat 901 was moved to a higher orbit for the maneuver and will then be taken back to a lower orbit and resume its interrupted communication mission.

Northrup-Grumman sees a future for satellite refueling and will launch another such mission soon.

Astronaut Edwin 'Buzz' Aldrin, courtesy NASA.

Chapter 7
Getting off to a Fast Start

Without a doubt we need to get off to a fast start. NASA's Artemis program to me looks too cumbersome and too complicated for what we need to do. Once we lay claim to the Moon we need to go there and never leave again. That's going to take a quick change to our priorities not just in space exploration but as a nation. Quite literally, we will need to put America first, and America fast!

The Space Force needs to be ready to go and it needs to be ready now! We will have those who seek to stop our plans as soon as they learn of them. That's not to say they aren't doing that very thing right now. But it does mean that once we state our intent to take the Moon they will put their own plans to thwart us in effect, and with a vengeance!

They will likely join forces and here I am talking about Communist China, Russia, and the smaller nations in their orbit. Conversely we have our own allies, nations we have worked with consistently over the years and although the Moon will belong to the U.S. we will likely invite them to join us there. This would be the United Kingdom, the European Union, Japan, India, and Israel at least. There may be others.

The United Nations will not be invited to come to the Moon. Their agenda remains much too anti-American.

This may give rise to another group of nations that I will call space want-to-be countries who will band together under the auspices of the UN.

Once American dominance is established on the Moon the Communist China – Russia axis will quickly turn their attention to gaining Mars and another space race will ensue. We must be ready to meet that challenge even as we coalesce our presence on the Moon. No time can be lost, no effort must be spared.

Not being able to leapfrog from the Moon to Mars will put a real cramp in the axis space powers move on Mars and give us a great leap forward. There is no reason why we should fear the loss of Mars if we do not dawdle or falter in an aggressive space exploration and colonization program. We need to relegate the axis space nations to smaller targets like asteroids and far away moons around Jupiter and Saturn.

Make no mistake, we will suffer setbacks, we will suffer casualties, and we will have times where we may doubt ourselves or our mission. But let me be clear.

The future of humanity is dependent upon our (American) ability to expand into space, to spread our people, people of all races into all the niches of the solar system. Yes, to boldly go and take residence in the many mansions that await our arrival. And yes, to eventually break loose from the restrictions of interstellar space travel and lay the foundations of that mythical United Federation of Planets dreamed up by science fiction writers and movie producers all those years ago. It is no doubt a dream worth pursuing for us and our children and our children's children.

Renewal of NASA Manned Spaceflight

NASA photo of Astronaut Doug Hurley in flight training.

Astronauts Doug Hurley and Bob Behnken will likely be the first NASA astronauts to fly aboard a commercial space vehicle, possibly as early as April or May of this year (2020). The two experienced spacemen will launch in SpaceX's Dragon Crew ship that just recently completed its crew abort test in an error free and virtually perfect mission.

Final training for the much anticipated launch that will signal NASA's return to manned spaceflight consists of renewed intensive training and practice runs of the launch procedures. The two veteran astronauts have worked together for the last twenty years and have developed a very real friendship.

NASA stated the two men have also been trained for retrieval from the sea after return and splashdown.

In a NASA video filmed in 2018 Hurley said "NASA has not done a flight-test program for a spaceship since the space shuttle. So you're talking late 70s, early 80s is the last time we kind of did this as an agency.

Some of it is kind of re-learning those techniques and those things that you need to make sure that you're watching out for."

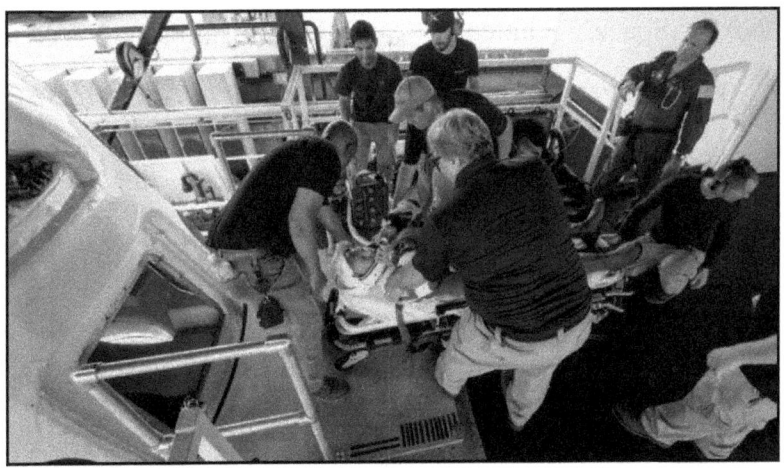

A retraction after splashdown training exercise. Photo courtesy of NASA.

NASA/SpaceX's Falcon 9 Crew Dragon Hangar. (NASA}

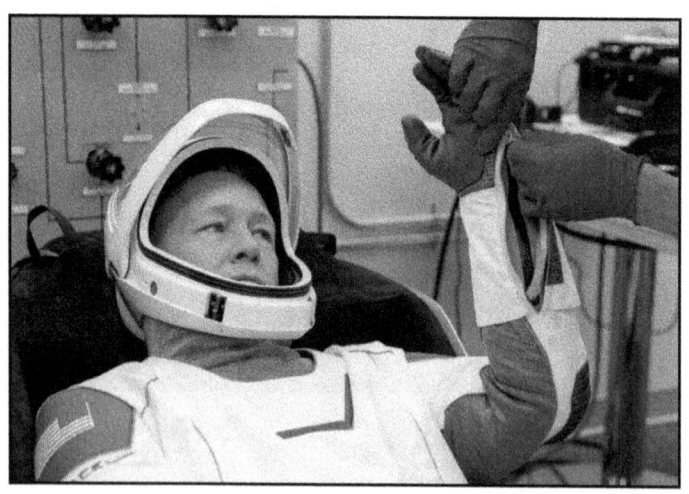

Getting fitted for the SpaceX spacesuit. NASA photo.

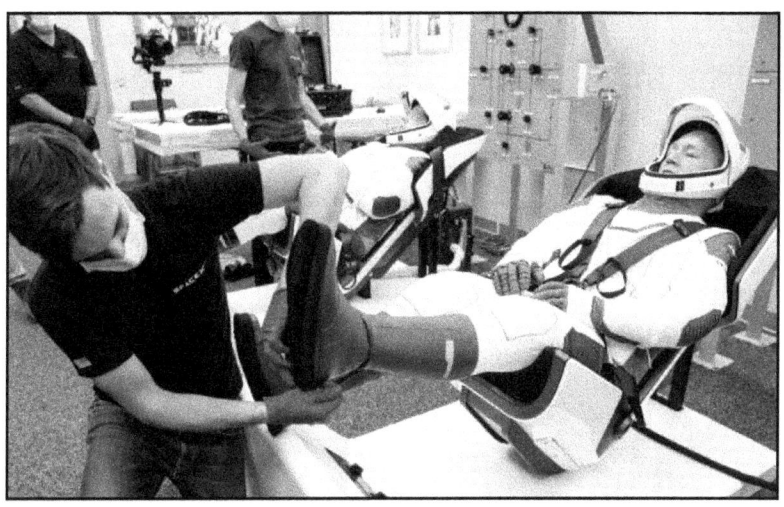

Crew Dragon dress rehearsal for SpaceX astronauts Behnken and Hurley. Photo courtesy NASA.

"People ask us about [the] commercialization of space, and I firmly believe that the more people we can get to go into space, the better off the planet's going to be." Hurley added.

In addition to the U.S. there are other nations who have launched astronauts on their own. These are of course Russia and China.

On January 1, 2020 India released the news that they plan to launch a manned mission to space in 2022. Although four astronauts have been selected, their names and backgrounds have not been released. If the plan is successful India will become the fourth nation to launch its own astronauts into space.

India has developed a spacecraft capable of human flight named Gaganyaan, Sanskrit for 'Sky Vehicle'. It is capable of taking up to three astronauts into space orbit for up to a week. The Indian astronauts are scheduled to start training in Russia in the near future.

It was also announced that its next robotic Moon mission, Chandrayaan-3 was approved and is already underway. Its anticipated launch is sometime in 2021. The attempt will include a lander and a rover similar to their last Chandrayaan-2 that partially failed and ended in a crash landing on the Moon.

New U.S. - Russia Space Race

SpaceX is expected to launch its first manned spaceflight in a matter of months. Once that happens, our reliance on Russian spacecraft to get our astronauts to the International Space Station will come to abrupt end.

This reportedly has triggered the Russian space agency to embark on building a new massive 246 foot tall rocket with ten engines dubbed the *Yenisei* that will consist of five stages. If true, it will no doubt be placed into direct competition with the U.S. to get manned missions to the Moon.

This rocket could lift huge loads to the Moon or to a space station orbiting the Moon. But frankly, I don't think Russia can afford such a program. The projected cost of developing and building such a spacecraft is estimated at no less than 22 billion dollars.

However the current crop of Russian rockets are not up to this level of mission. So if they really want to compete with Communist China and the U.S. in the new race to the Moon, this is the way they could do it. A big problem is that it's not expected to be available, even if everything goes well, until 2028. In the real world of space exploration, that will be way too late.

The U.S. Space Launch System is scheduled for its first launch this year, and when operational, it is expected to be able to carry payloads of up to 130 tons, much more than Yenisei's projected payload capacity of 80 tons.

It appears to me that Russia has already lost this space race before it even begins. It's my belief this is merely the Russian's attempt to get us to let them join us on the Moon. If so, we already have leverage over them.

Upcoming Space Missions for 2020

As mentioned earlier, both SpaceX and Boeing are expected to launch astronauts during this year. And, although they are likely somewhat behind, there may be an outside chance that Blue Origin may put a man or woman into space before the end of the year. All three company's spacecraft are still being tested to insure they are safe for human transport. One or all of these space agencies may launch astronauts before I complete this book and if so I will, of course, include that information.

The Solar Orbiter, sponsored by both NASA and ESA is to launch and take up its orbit around the Sun in February. The spacecraft's seven year mission is to get very close to study the Sun, so close that it will be stationed inside the very warm orbit of Mercury.

Mars will approach close to Earth in July and the favorable position will result in the launch of no less than four Mars missions.

NASA will begin the launch series with the 'Mars 2020 Rover' on July 17. This is to be followed shortly by the joint Russian and ESA launch of a rover to Mars on July 25. Not far behind will be a United Arab Emirates' mission to be launched by Japan and dubbed the 'Hope Mars Mission'. Communist China is also scheduled to launch a Mars rover.

Communist China will likely launch it next lunar mission titled Chang'e – 5 in 2020. This time their intent is to land on the Moon, collect a sample of the surface and return it to Earth for analysis. Not only will this mission have the prestige of joining the U.S. and the Soviet Union as the only other countries to secure a Moon sample, but is also good practice in landing a manned lander and bringing it back to the Earth. That means they will have the edge on us in creating a direct manned express to the Moon. A direct manned express is exactly what we need to do. China may steal a march on us.

The mission calls for the lander to set down near Mons Rumker a mountain in the lunar sector called Oceanus Procellarum.

The U.S. Space Force will launch the sixth classified mission of the X-37B space shuttle, also known by the innocuous name of the Orbital Test

Vehicle. The OTV has performed a mission that required it to be in space for a record breaking 780 day spaceflight. That is a whopping two years and fifty days!

The fact sheet on this secret spacecraft states the mission is for testing "Reusable spacecraft technologies for America's future in space and operating experiments which can be returned to, and examined, on Earth."

SpaceX Requests Early Starship Launch

SpaceX would like to launch the Mark 3 prototype Starship as early as March of this year for a test flight up to 12+ miles high. If approved, the Starship will take off from the company's launch complex in Boca Chica, Texas. The company expects to launch the ship a second time in 2020 to an altitude of 62 miles.

The Starship was unveiled in September of last year and is the next generation of launch vehicle for lifting people and cargo both to the Moon and later to Mars. The experimental launch is to give the ground crew the opportunity to train by communicating to the ship in flight and monitoring its operation.

The data gathered will be shared with the U.S. Air Force, NASA, and other appropriate government agencies.

SpaceX to Speed Starship Production
A Cautionary Note

Elon Musk's space company expects to get to a stable Starship design that will allow it to begin a high rate of its production. Like anything else that is repeated over and over the price of manufacturing that item will go down. That would allow mass selling around the world to take place.

Although that might make money for SpaceX, would the U.S. want unregulated spaceships sold to every country that might want to, for example, go to the Moon? I don't think the U.S. can allow the willy-nilly selling of large rockets that can attain orbit and make a hell of a big bomb.

The company is hiring more personnel and will seek to ramp up the design and production of the large reusable spacecraft. Could this become something that may need government oversight? I would think so.

Space Tourism 2020

The space tourism company Virgin Galactic is aiming to begin space tourist flights to the edge of space with paying tourists on board.

The spaceship planes will launch from Virgin Galactic's New Mexico base dubbed 'Spaceport America'. These flights will be suborbital and designed to give the passengers a taste of space flight and exposure to momentary weightlessness.

For this trip of a lifetime, each tourist will be tagged for the 250K price of admission.

The VSS Unity will hold six paying passengers and two pilots. That gives each flight a cool $1.5 million in profit. Each round trip flight is approximately two hours in length. The flights are tentatively scheduled to begin mid-year.

Not to be outdone, Blue Origins is also determined to join the tourist space race using their New Shepard spacecraft. However, Blue Origins seems to me to be well behind Virgin Galactic. I make the prediction that they will not begin tourist flights in 2020. And if they do, it will be well toward the end of the year.

A Clock Work Orange on Venus

Due to the massive atmospheric pressure and extreme heat on Venus, only the most robust form of a rover could possibly survive on its surface.

Recognizing those facts, NASA has sent out a call for the development of a clock work sensor mechanism that could not only survive on the hostile world, but be able to direct a rover to travel around on it.

The "Exploring Hell: Avoiding Obstacles on a Clockwork Rover" initiative requests the public's input on creating such a robust mechanical sensor.

Every mission's equipment that has reached the surface of our so called 'twin' planet has quickly succumbed to the extremes of the planet's atmosphere.

The last mission to land on Venus was the 1985 Russian Vega-2 which didn't fare any better than the earlier attempts. But that has not stopped NASA/JPL from planning to launch the Automaton Rover for Extreme Environments (AREE) program. The concept has also launched a hybrid kind of engineer dubbed a 'mechatronics specialist' at the Jet Propulsion Laboratory.

Jonathan Sauder is one such specialist, in fact he is the senior specialist on the program and stated:

"Earth and Venus are basically sibling planets, but Venus took a turn at one point and became inhospitable to life as we know it. By getting on the ground and exploring Venus, we can understand what caused Earth and Venus to diverge on wildly different paths and can explore a foreign world right in our own backyard."

AREE relies on the winds of Venus to move it around while the clock work sensor picks up and records the data. Got an idea? Get it to the JPL.

The near side of the Moon, the side we are most familiar with. (NASA)

Chapter 8
Claiming the Moon

In July of 1969 I, like millions of other people around the world watched as the U.S. space ship *Apollo 11*, under the command of Astronaut Neil Armstrong, landed on the surface of the Moon. It was one of the most momentous events in the history of humanity.

Other *Apollo* landings were to take place in the future but none of them as important as this first landing. The last time a manned space craft landed upon the Moon was *Apollo 17* on December 7, 1972. Since that ship took off to return to Earth less than two weeks later, no human has set foot again upon the Moon. These facts give the United States a clear claim to some, if not all resources on Luna.

Why would we want to lay claim to the Moon? I believe and I am not alone in that belief, that having rights to the resources of the Moon such as Helium-3 will be worth trillions, perhaps even more.

Helium-3 (H3) is a light non-radioactive stable isotope of Helium discovered in 1939. It is a potential gas fuel that can be used in powering nuclear reactors. There is very little Helium-3 on Earth, but it appears to be available in large quantities on the Moon.

Water, I don't think I have to explain too much about why having a source of water on the Moon is priceless for any human operation on Earth's satellite. In addition to the obvious uses humans have for water ($H2O$) is the fact that it can be refined into both Oxygen and Hydrogen. The need for Oxygen goes without saying and the Hydrogen makes an excellent rocket fuel. Other valuable resources are gold, platinum, and rare earth metals.

Yes, the composition of the Moon is so similar to Earth that it contains earth metals.

Other valuable minerals include silicone, iron, magnesium, calcium, aluminum, manganese, and titanium. These minerals will be essential to any Moon settlement.

Occupying the Moon, building settlements, and protecting it all with the Space Force will insure that the United States of America will have the upper hand in the development of space. It will assure our future access to the endless wealth of the Moon itself and eventually the entire planet of Mars!

You only have to take a peek back in history when Spain, followed closely by Portugal, lay claim to the western hemisphere of Earth. It guaranteed their dominance and wealth for centuries. Laying claim will do the same for us, but it will guarantee our dominance and wealth virtually forever.

I think I have demonstrated that the Moon is worth the cost of returning to its surface and laying a claim to it. Now, how can we do that?

The last territorial acquisition by the United States was the Hawaiian Islands annexed by President William McKinley in 1898. The Stars and Stripes had flown on the islands since 1893 when the U.S. Marines stormed ashore and the Queen of Hawaii was deposed. It was a messy affair. Later, the Congress voted a Joint Resolution of only one page in length annexing Hawaii. But annexing the Moon shouldn't be messy but without a doubt somewhat complicated, for the following reasons.

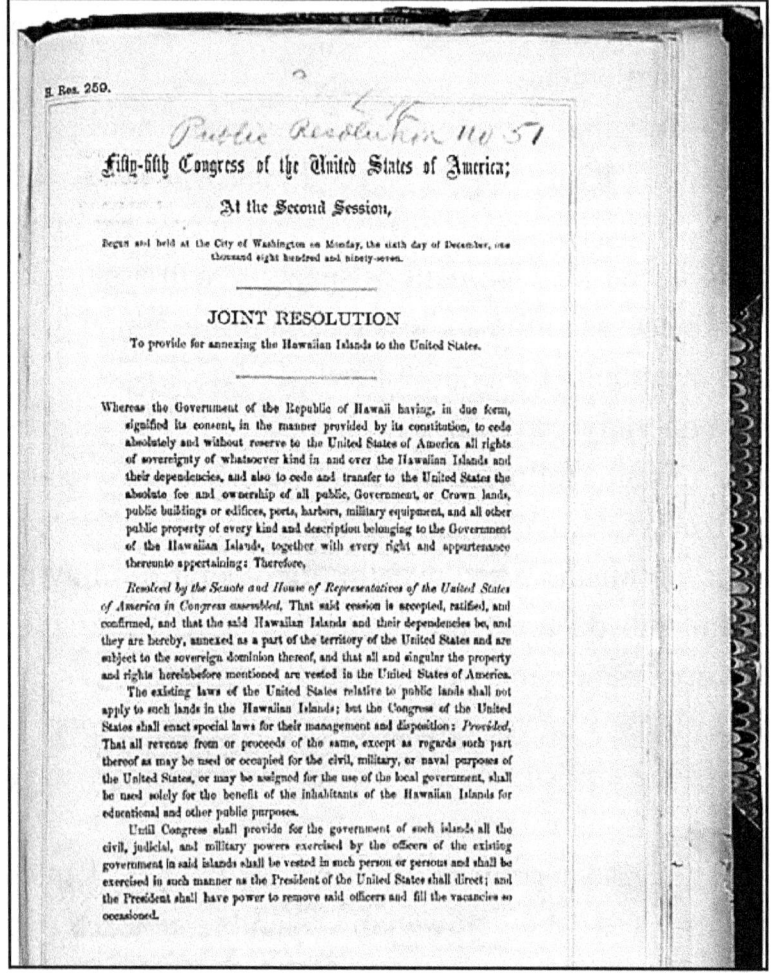

That's it, one page to annex the Hawaiian Islands, initially a territory and now a state worth trillions of dollars today.

There are no indigenous people on the Moon. In fact, the closest people to approach that status are the twelve U.S. astronauts, Americans all, who have walked on the surface of the Moon and the twelve others who have orbited the Moon.

There are no other national claimants that even approach the use the U.S. has made of the Moon, other than using the light of the Moon at night, on Earth.

Although I understand that the U.S. flag we left on the Moon has now been bleached white by the sun, it remains our flag. And it has been on the Moon for the past fifty plus years!

Recent NASA photographs taken by the Lunar Reconnaissance cameras of the six Apollo landing sites have demonstrated that the flags planted by our astronauts are still flying (figuratively speaking as there is no wind) on the Moon's surface.

Land Claims in the U.S.

A land claim is a declaration of control over areas of property including the very important claim over water rights. Claims can stem from a number of variables that render different levels of merit or legitimacy of the claim. The four variables are:

Claim without any actual action on the surface. Claim with moveable property on the surface. Claim with the claimant visiting the surface. Claim with the claimant living on the surface.

Each of these levels of claim makes a stronger legal case for those who can prove them. At this time, only the United States has attained all four levels of a legal claim.

However, there are countries gearing up space programs even as I write these words to push to the Moon. The most aggressive of these is Communist China and Russia, although China seems poised to get there sooner than any space faring nation except the United States.

Mining Claims in the U.S.

Merriam-Webster's definition of a mining claim is as follows: a tract of land having access to a vein or lode of valuable minerals supposed to exist below and definitely located on its surface by a miner with the right to occupy and mine in the manner and under the conditions prescribed by law usually involving discovery and the filing of legal notice.

The U.S. Bureau of Land Management's definition is: A mining claim is a parcel of land for which the claimant has asserted a right of possession and the right to develop and extract a discovered, valuable, mineral deposit. This right does not include exclusive surface rights (see Public Law 84-167).

I would hazard a guess that if the federal government of the United States claimed property on the Moon, the public law cited above could be easily overruled.

This definition by Wikipedia gives a little more background: A mining claim is the claim of the right to extract minerals from a tract of public land. In the United States, the practice began with the California gold rush of 1849. In the absence of effective government, the miners in each new mining camp made up their own rules, and chose to essentially adopt Mexican mining law then in effect in California. The Mexican law gave the right to mine to the first one to discover the mineral deposit and begin mining it. The area that could be claimed by one person was limited to that which could be mined by a single individual or a small group.

Staking a Claim

Again, Wikipedia says: Staking a claim involves first the discovery of a valuable mineral in quantities that a "prudent man" (the Prudent Man Rule) would invest time and expenses to recover. Next, marking the claim boundaries, typically with wooden posts or capped steel posts, which must be four feet tall, or stone cairns, which must be three feet tall. Finally, filing a claim with the land management agency (I assume that would be NASA in the name of the American people).

There are four main types of mining claims:

1. Placer (minerals free of the local bedrock, and deposited in benches or streams)
2. Lode (minerals in place in the mother rock),
3. Tunnel (a location for a proposed tunnel which claims all veins discovered during the driving of it)
4. Millsite (a maximum five acre site for processing ore)

A patented claim is one for which the U.S. federal government has issued a patent (deed). To obtain a patent, the owner of a mining claim must prove to the federal government that the claim contains locatable minerals that can be extracted at a profit. A patented claim can be used for any purpose desired by the owner, just like any other real estate. However, Congress has ceased funding for the patenting process, so at this time a claim cannot be patented.

How to Stake a Moon Claim

First, you have to be a U.S. citizen. Then, and this is a big step, you have to get to the Moon and physically stake your claim. Americans who wanted to stake a claim in Hawaii or Alaska when they became territories had to go there too. So, how do you get to the Moon?

If you are a billionaire I suggest you talk to Elon Musk and hitch a ride to the Moon. You can likely afford the ticket. Or, you could buy a spaceship. Space-X, Blue Origin, and others have spaceships you can likely purchase and there are rocket launch facilities you can rent or lease. Don't forget to take your spacesuit and a shove.

If you're not a billionaire, are young, smart in math and science, possessing a good deal of courage, fortitude, and the rest of the 'right stuff' apply at NASA to be an astronaut.

If you feel you have the right stuff to become a military officer, non-commissioned officer, or foot soldier, volunteer for the Space Force.

If you are neither of these but are at least young, sign up to migrate to the Moon as a technician. You will have to have a skill useful on the Moon. Fortunately just about every job we have on Earth will eventually be on the Moon as well. Apply at NASA.

Outer Space International Treaties

You may be thinking aren't we bound by international treaty regarding claims in outer space? Well, there is a treaty and it's titled:

Treaty on Principles Governing the <u>Activities</u> of States in the Exploration and Use of Outer Space, including the Moon and Other Celestial Bodies.

The treaty consists of this preamble and has seventeen articles. If you have an interest in the articles themselves there are ways to access them either on line or in written form from most libraries. The preamble reads as follows:

The States Parties to this Treaty,
Inspired by the great prospects opening up before mankind as a result of mans entry into outer space,
Recognizing the common interest of all mankind in the progress of the exploration and use of outer space for peaceful purposes,
Believing that the exploration and use of outer space should be carried on for the benefit of all peoples irrespective of the degree of their economic or scientific development,
Desiring to contribute to broad international co-operation in the scientific as well as the legal aspects of the exploration and use of outer space for peaceful purposes,
Believing that such co-operation will contribute to the development of mutual understanding and to the strengthening of friendly relations between States and peoples,
Recalling resolution 1962 (XVIII), entitled "Declaration of Legal Principles Governing the

Activities of States in the Exploration and Use of Outer Space," which was adopted unanimously by the United Nations General Assembly on 13 December 1963,

Recalling resolution 1884 (XVIII), calling upon States to refrain from placing in orbit around the Earth any objects carrying nuclear weapons or any other kinds of weapons of mass destruction or from installing such weapons on celestial bodies, which was adopted unanimously by the United Nations General Assembly on 17 October 1963,

Taking account of United Nations General Assembly resolution 110 (II) of 3 November 1947, which condemned propaganda designed or likely to provoke or encourage any threat to the peace, breach of the peace or act of aggression, and considering that the aforementioned resolution is applicable to outer space,

Convinced that a Treaty on Principles Governing the Activities of States in the Exploration and Use of Outer Space, including the Moon and Other Celestial Bodies, will further the Purposes and Principles of the Charter of the United Nations,

Have agreed on the following: (the seventeen articles follow).

We need to TRUMP this treaty (yes, pun intended), so I certainly hope we have the U.S. Attorney General working on a way to withdraw from this ill-conceived and obsolete treaty and restore our sovereign rights to lay claim to property in outer space. I have little doubt but that President Trump will support such a withdrawal.

Ipso Facto

Most signatories of this treaty either signed or ratified this or did both. Communist China did neither of these. They performed an accession of the treaty. What does that mean?

There seems to be a number of legal definitions of accession, all of them rather slippery. It appears to actually mean that an entity, whether an individual, a group of individuals or a nation, agrees substantially to the form (of the document). When was the last time you signed a contract that only required you to substantially agree with what was written? I'll hazard a guess and say never. Yet that is all that Communist China agreed to. Why?

It should be apparent, it's because they never intended to abide by the treaty. They are notorious for signing treaties and not abiding by them. Just look at what they are doing in the South China Sea. Here are the titles of some analysis and comments on Communist China ignoring international treaties.

China versus Vietnam: An Analysis of the Competing Claims in the South China Sea.

China's new '10 – Dash line Map' Eats into Philippine Territory

PH (Philippines) Protest China '9 – line' [and] *Spratly* [islands] *Claim*

U.S. Navy: Beijing is creating a great wall of sand in the South China Sea

China's Nine Dash Line faces renewed assault

    These are only a few articles and papers protesting or drawing attention to Communist China's ignoring of international treaties. Too let China know this is unacceptable, the U.S. Navy regularly sends warship patrols around and through the Spratly Islands ignoring their illegal claims of sovereignty.

    So, they are not bound by the international treaty of 1967 and oh yeah, they are our greatest competition for laying claim to the Moon. They will not abide by the international treaties in outer space either.

    Now I'll use my limited lawyer jargon and say *ipso facto* (by that act) of China's mere accession of the treaty, the United States are not bound to abide by it either.

    The U.S. position has been and continues to be that it will recognize Moon resources developed by our nationals, as their property!

The Space Act of 2015

The Commercial Space Launch Competitiveness Act, sometimes referred to as the 'Spurring Private Aerospace Competitiveness and Entrepreneurship (SPACE) Act of 2015', is an update by the United States government of its commercial space use policy legislated in 2015.

Enacted by the 114th United States Congress sometimes it is simply called the 'Space Act of 2015'.

In November of 2015, Congress passed this act which states that private American companies own any resources they collect from celestial bodies such as asteroids or the Moon.

Less than a year later on August 3, 2016 the United States Federal Aviation Administration (FAA) authorized the first private lunar mission. This became a fact when it permitted Moon Express, Inc. also known as MoonEx to travel beyond Earth's orbit, navigate to the Moon, land upon its surface, and conduct both lunar exploration and discovery of the Moon's resources.

MoonEx openly claimed this was a breakthrough in U.S. policy that marked the beginning of a new era of ongoing commercial activity on the Moon to unlock its valuable resources. This assertion seems hard to argue with.

My take on this whole issue is that someone, and here I am speaking about the government of the United States, needs to clear away any ambiguity about claims in outer space. The way to do that is, as I have said before, for us to withdraw from the treaty on outer space and to lay claim to the Moon.

This could initially be done by a presidential decree followed up by an act of congress. All we need is the courage to do it.

No doubt there will be an outcry around the world with Communist China crying the most. They will quickly follow-up with a call for an emergency U. N. meeting.

However, the truth is that we have the upper hand, we have the justification to lay such a claim, and we must do it now!

The Trailing side of the Moon, half way between the near and far sides. Photograph courtesy of NASA

Chapter 9
Justifying Our Claim

First I have to point out that it should be obvious to anyone who even has a superficial knowledge of the outer space that the United States is and has dominated space exploration since the 1960s. We are the premier nation on Earth in space science and soon will also lead in the economic development as well.

And here I have to give a nod to the patriotic and farseeing businessmen and corporations who are contributing quite favorably to this economic growth. Among them of course are Elon Musk of SpaceX; Jeff Bezos of Blue Origin, Boeing, Lockheed-Martin, Northrop Grumman, Astrobotic (robot lunar landers), Axiom (commercial space stations), and many others launching satellites and developing a tourist industry. These business firms alongside NASA are the promise of the future. Let's insure they get the first crack at settling on the Moon and conducting American business.

Initially, our claim is supported by the many firsts we have accomplished while voyaging to, landing upon, and occupying the Moon. What follows is a list of these astonishing Moon firsts.

Apollo 8 December 1968, the first crewed space craft to orbit the moon.

Apollo 11 First human landing on the Moon July 1969.

Apollo 12 First targeted pilot landing on the Moon November 1969.

Apollo 15 First use of a vehicle, the lunar rover, on the Moon.

Apollo 16 was the tenth crewed mission in the United States Apollo space program, the fifth and second-to-last to land on the Moon, and the second to land in the lunar highlands. The second of the so-called "J missions,"[53] it was crewed by Commander John Young, Lunar Module Pilot Charles Duke and Command Module Pilot Ken Mattingly.

Launched from the Kennedy Space Center in Florida at 12:54 PM EST on April 16, 1972, the mission lasted 11 days, 1 hour, and 51 minutes, and concluded at 2:45 PM EST on April 27.

Young and Duke spent 71 hours—just under three days—on the lunar surface, they conducted three extra-vehicular activities or moon walks, totaling 20 hours and 14 minutes. The pair drove the Lunar Roving Vehicle (LRV), the second produced and used on the Moon, for 26.7 kilometers (16.6 mi). On the surface, Young and Duke collected 95.8 kilograms (211 lbs.) of lunar samples for return to Earth, while Command Module Pilot Ken Mattingly orbited in the command and service module (CSM) above to perform observations. Mattingly, staying with the command module, spent 126 hours and 64 revolutions in lunar orbit.[6] After Young and Duke rejoined Mattingly in lunar orbit,

---

[53] J missions were longer three-day stays using an extended lunar module, with three lunar extravehicular activities and a Lunar Roving Vehicle.

the crew released a sub satellite from the service module (SM). During the return trip to Earth Mattingly performed a one-hour spacewalk to retrieve several film cassettes from the exterior of the service module.

Apollo 16's landing spot in the highlands was chosen to allow the astronauts to gather geologically older lunar material than the samples obtained in three of the first four Moon landings, which were in or near lunar maria (Apollo 14 landed in the Fra Mauro Highlands). Samples from the Descartes Formation and the Cayley Formation disproved a hypothesis that the formations were volcanic in origin

Apollo 17 (December 7-19, 1972) was the final mission of NASA's Apollo program; it remains the most recent time humans have travelled beyond Earth orbit. Its crew consisted of Commander Eugene Cernan, Lunar Module Pilot Harrison Schmitt, and Command Module Pilot Ronald Evans. It carried a biological experiment containing five mice.

Launched at 12:33 a.m. Eastern Standard Time (EST) on December 7, 1972, Apollo 17 was a "J-type mission" that included three days on the lunar surface, extended scientific capability, and the use of the third Lunar Roving Vehicle (LRV).

Cernan and Schmitt landed in the Taurus–Littrow valley and completed three moon walks, taking lunar samples and deploying scientific instruments.

The landing site was chosen to further the mission's main goals: to sample lunar highland material older than Mare Imbrium, and to investigate the possibility of relatively recent volcanic activity. Evans remained in lunar orbit in the command and service module (CSM), taking scientific measurements and photographs. Astronauts Cernan, Evans, Schmitt, and the mice returned to Earth on December 19.

The Apollo 17 mission included the first night launch of a U.S. human spaceflight and the final crewed launch of a Saturn V rocket. It was also the final use of Apollo hardware for its original purpose.

The mission broke several crewed spaceflight records: the longest Moon landing, longest total extra-vehicular activities (moon walks), largest lunar sample, and longest time in lunar orbit and, at 75, most lunar orbits.

These missions met all four criteria for a land claim under U.S. law. So, what has to happen to lay the actual claim?

First, we repudiate and withdraw from any international treaties regarding outer space. Second, we lay our claim before the U.S. federal court system. Third, we return to the Moon and establish not only military bases and construction bases, but support bases as well; we also establish a colony and mount Space Force patrols because there are powers on Earth who will not be happy about this. Communist China comes to mind for sure. Then there is Russia, the heir to the old Soviet Union space program, proud but seeming to lack the funding to compete for the Moon. India too is very interested in the Moon and its latest space endeavor is to get close to the water sites of the lunar surface. Then there is the European Union, a studious yet seemingly cooperative competitor. They could likely be our allies in our Moon endeavors.

Space buoys broadcasting in every Earth language will warn approaching space craft that they must obey the instructions of our space craft approach and control center and the orders of Space Force officers. These things must be done immediately.

Prince Rupert's Land

In 1670 the largest commercial land claim was granted by the British Crown to the Hudson Bay Company. This is certainly a precedent of a huge land claim from a government to a commercial enterprise. This royal charter gave sole proprietary rights to nearly five million square miles of what later became a large part of Canada and some small areas of what became the United States.

This claim was granted to the Hudson Bay Company, a fur trading organization. For the next two hundred years this company had the sole rights to virtually everything within this claim. Finally in

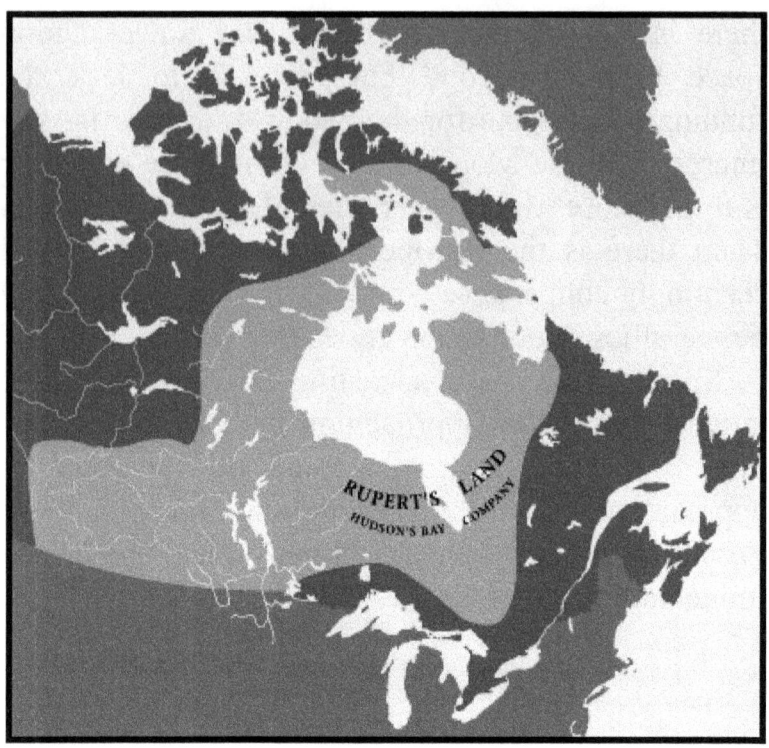

1868 the land was sold to the government of Canada, except those areas inside the United States.

If we wanted to claim the Moon we could go this route, have a commercial company lay the initial claim and present it to the government for approval. The government could deny claim and claim the Moon in the interests of the United States and so stop any further commercial claims.

However, we don't really have to take this step, because it was the government, in the form of NASA that first landed upon and occupied the Moon with our citizens. And, it remains the only Earth government ever to have done so.

Columbus' Claim

Like most modern historians, I admit that Columbus' claim to the New World was a catastrophe for the Native Peoples of the Americas. His greed for gold caused great suffering to millions of people for centuries.

However, I will point out again, there are no indigenous people on the Moon. Indeed, the only people to have traveled to and/or trod upon the surface of the Moon are twenty-seven American men of mostly European extraction. Twelve of them landed upon the Moon's surface and another fifteen of them orbited the Moon. And yes, that gives the United States of America, the only legitimate national claim, by right of exploration and occupation, to the physical confines and the adjacent space and orbiting lanes, of the Moon. This claim is in keeping with history and maritime law.

By substituting the word explorers for salvagers and the word discovery for the word salvage these five statements taken from maritime law in fixing an award of value is appropriate in my view.

1. The labor extended by the explorers in rendering the discovery services;
2. The promptitude, skill, and energy displayed in rendering the discover services;
3. The value of the property employed by the explorers in rendering the services and the danger to which such property is exposed;
4. The risks incurred by explorers in securing the property.
5. The value of the property secured.

Our astronauts did all of this and even more because they greatly added to the knowledge of the Moon and space through their actions.

There are, of course, no people on the Moon, our claim will be nothing like those of the past. We must pursue this course of action immediately!

The Far side of the Moon.

Chapter 10
Why We Must Lay Our Claim Now

Mission Clementine

The possibility of ice on the moon was suggested in 1961. The suggestion stated that water vapor escaping the crust of the primitive moon or deposited on the lunar surface by comets and asteroid impacts may have collected in permanently shadowed craters. These craters are referred to as cold traps and they do in fact exist at the lunar poles. The resulting ice has remained stable in these cold traps over geological time.

The cold trap area at the South Pole is more extensive than at the North Pole, it was expected to retain more trapped ice and so it has proved.

Clementine was a joint project between the Strategic Defense Initiative Program, begun by President Reagan, and NASA. The objective of the mission was to test sensors and spacecraft components under extended exposure to the space environment and to make scientific observations of the Moon and the near-Earth asteroid 1620 Geographos.

The observations included imaging at various wavelengths including ultraviolet and infrared, laser ranging altimetry[54], and charged particle measurements. These observations were originally for the purposes of assessing the surface mineralogy of the Moon and Geographos, obtaining lunar altimetry from 60N to 60S latitude, and determining the size, shape, rotational characteristics, surface properties, and cratering statistics of Geographos.

Clementine was launched on 25 January 1994 at 16:34 UTC (12:34 PM EDT) from Vandenberg Air Force Base aboard a Titan IIG rocket. After two Earth flybys, lunar insertion was achieved on February 21. Lunar mapping took place over approximately two months, in two parts. The first

---

[54] The measurement of altitudes from a surface such as hills or valleys.

part consisted of a 5 hour elliptical polar orbit with a perilune[55] of about 400 km at 28 degrees south latitude. After one month of mapping the orbit was rotated to a perilune of 29 degrees north latitude, where it remained for one more month. This allowed global imaging as well as altimetry coverage from 60 degrees south to 60 degrees north.

After leaving lunar orbit, a malfunction in one of the on-board computers on May 7 at 14:39 UTC (9:39 AM EST) caused a thruster to fire until it had used up all of its fuel, leaving the spacecraft spinning at about 80 RPM[56] with no spin control. This made the planned continuation of the mission, a flyby of the near-Earth asteroid Geographos[57], impossible. The spacecraft remained in geocentric orbit and continued testing the spacecraft components until the end of its mission.

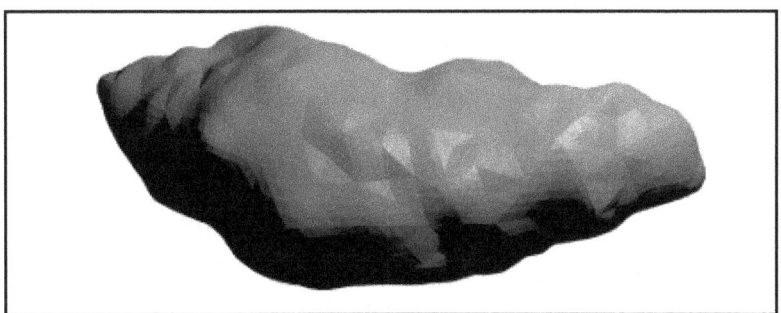

A NASA model image of Geographos.

---

[55] The point at which a spacecraft in lunar orbit is closest to the moon.
[56] Revolutions per Minute.
[57] 1620 Geographos, provisional designation 1951 RA, is an elongated, stone type asteroid, a near-Earth object and potentially a hazardous asteroid of the Apollo group, with a diameter of about 1.6 miles.

Bistatic Radar Experiment

The Bistatic Radar Experiment improvised during the mission, was designed to look for evidence of lunar water at the Moon's poles. Radio signals from *Clementine* probe's transmitter were directed towards the Moon's north and south polar regions and their reflections detected by Deep Space Network receivers on Earth. Analysis of the magnitude and polarization of the reflected signals suggested the presence of volatile ices, interpreted as including water ice, in the Moon's surface soils. A possible ice deposit equivalent to a sizeable lake was announced. However, later studies made using the Arecibo radio telescope showed reflection patterns even from areas not in permanent shadow (and in which such volatiles cannot persist), leading to suggestions that *Clementine*'s results had been misinterpreted and were probably due to other factors such as surface roughness.

NASA announced on March 5, 1998, that data obtained from *Clementine* indicated that there is enough water in polar craters of the Moon to support a human colony and a space craft fueling station.

This establishes that it was the United States that initially discovered water on the Moon and therefore puts our claim to this invaluable resource as premiere.

NASA's LCROSS Mission

This little known mission may end up being one of the most important in the annals of space exploration because of what it found. It found and confirmed water on the Moon, a source of oxygen for breathing, water for drinking, and hydrogen to fuel our spacecraft. On a world that has little of any of them, its importance cannot be dismissed.

During NASA's preparation for launching the Lunar Reconnaissance Orbiter it was found necessary to change the spacecraft's launch vehicle from the scheduled Delta II rocket to a more powerful vehicle. Doing that freed up a good deal of payload space and it was decided to combine two missions into one.

The second mission was the Lunar Crater Observation and Sensing Satellite or LCROSS. Its mission was to determine the nature of hydrogen detected at the polar regions of the Moon. This mission was to further explore the presence of water ice in a permanently shadowed crater near a lunar polar region.

LRO/LCROSS drawing courtesy of NASA

The dual mission of the LRO and LCROSS were launched on June 18, 2009, the vanguard of NASA's return to the Moon.

LCROSS was to collect and relay data from an impact and debris plume resulting from the acceleration and crash of the vehicle's spent Centaur upper stage and the data collecting spacecraft behind it, and finally striking the crater Cabeus near the south pole of the Moon. The Centaur crashed on October 9, 2009, and the following sensor spacecraft traveled through the plume of dust, collected and relayed the data, and crashed six minutes later. It was quite a well thought out and ingenious feat of space exploration!

On 13 November 2009, NASA reported that evidence from analysis showed water was present in both the high-angle vapor plume and the ejected material resulting from the LCROSS Centaur impact. Additional confirmation came from an emission in the ultraviolet spectrum that was attributed to hydroxyl fragments, a product from the break-up of water by sunlight.

2019 analysis shows that each pole's cold traps contain an estimated ONE BILLION METRIC TONS OF WATER EACH!

This water ice on the Moon is priceless. No permanent human base or colony can endure on the

Moon without it. It is the main reason we need to claim the Moon now, before Communist China or Russia, or any other space faring nation does it first.

Now how silly does President Trump look for establishing the U.S. Space Force? Why, not at all, in fact he looks like a genius. Any base or colony on the Moon placed by the U.S. must include a USSF base at each pole as a minimum!

Possible Claim Challengers

Who are the space faring nations of Earth? The dictionary definition of a space faring nation is: Any nation with the capacity to travel in space.

A similar but not the same definition of space capability is: 1.The ability of a space asset to accomplish a mission. 2. The ability of a terrestrial-based asset to accomplish a mission in space (e.g., a ground-based or airborne laser capable of negating a satellite).

The US accounts for approximately one-third of the operational spacecraft currently in orbit around Earth. The Apollo moon-landing missions, the Skylab space station, Space Shuttle, International Space Station (ISS), Mars Exploration Rover – Opportunity, and Mars rover Curiosity are the cornerstones of the nation's space program.

The nation launched its first satellite into space in February 1958 and currently operates a large fleet of communications, electronic intelligence, missile detection, weather, technology, navigation, and surveillance satellites. The national space exploration efforts of the country are led by the National Aeronautics and Space Administration (NASA).

China owns and manages the second largest fleet of spacecraft in orbit, currently operating several constellations of navigation satellites, remote sensing satellites, communication satellites, surveillance and spacecraft. Communist China is one of three nations with the capability to recover satellites and conduct a manned space flight.

China's major missions include the Tiangong-1 space station, Shenzhou manned space flight program and the Chinese Lunar Exploration program (CLEP). The Chinese National Space Administration (CNSA) handles the planning and development of its national space programs, while state-owned China Aerospace Science and Technology Corporation (CASC) is the prime contractor responsible for the design and development of launch vehicles and satellites as well as commercial launch services.

The origins of the impressive Russian space program can be traced back to 1957 when the world's first artificial satellite Sputnik 1 was launched by the Soviet Union. The country now operates the third largest fleet of spacecraft including communications, meteorological and reconnaissance satellites.

Projects include the Soyuz manned spacecraft, Salyut 1 space station and Lunokhod 1 space rover. The Russian Federal Space Agency (ROSKOSMOS) supervises civilian space activities, whereas the

Russian Space Forces (VKS) handles defense satellite launches and military flight control assets.

Japan launched its first satellite Osumi into space in February 1970, becoming the fourth nation after the USSR, the US and France to possess indigenous satellite launch capability. It currently operates a fleet of communications, meteorological, earth observation and astronomical observation satellites.

Notable Japanese space programs are the Japanese Experiment Module (KIBO)-ISS, H-II Transfer Vehicle KOUNOTORI5 (HTV5) and H-II launch vehicle. The national aerospace research and development activities are controlled by the Japan Aerospace Exploration Agency (JAXA).

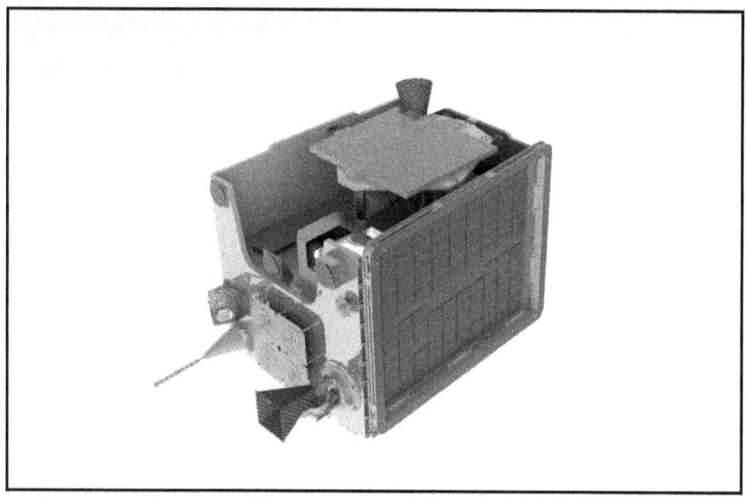

The United Kingdom launched its first satellite, Ariel 1, in 1962, making it the third nation after the USSR and the US to launch artificial satellites into orbit. It presently operates a large number of satellites including civil and military communications satellites, earth observation satellites, and scientific and exploration spacecraft.

The UK is one of the largest monetary contributors to the European Space Agency (ESA) and participates in advanced science and exploration missions such as Bepi Colombo, Euclid and ExoMars Rover carried out by the ESA. The United Kingdom Space Agency (UKSA) is responsible for the implementation of the national civil space program.

India has launched more than 80 spacecraft since its maiden satellite launch in 1975. The nation's space research activities are controlled by state-owned Indian Space Research Organization (ISRO). India currently operates INSAT and GSAT series communication satellites, earth observation satellites, and IRNSS series navigational satellites.

Dual-use satellites such as TES and Cartosat serve both civilian and military applications, whereas India's first dedicated defense satellite GSAT-7 serves the military. India also executed the Mars Orbiter Mission (MOM) at a cost of $75m, spending approximately one-tenth of what NASA did on the MAVEN Mars mission.

On July 22, 2019 India launched a spacecraft intended to closely explore water deposits near the South Pole of the Moon.

Dubbed the Chandrayan, Sanskrit for moon craft was to land at the lunar South Pole and physically explore the area using a rover.

India's 2008 moon mission, its first, carried U.S. scientific instruments that confirmed the presence of water at the poles. The fact that the U.S. is working to send a manned mission to the moon's South Pole by 2024 underscores the importance of this development.

The Indian spacecraft carried an orbiter, a moon lander and a rover designed to move around on the lunar surface for fourteen Earth days.

On September 6, 2019 India's mission control lost radio contact with the Vikram moon lander during its landing procedure. The mission had failed.

The orbiter was designed to remain in orbit for approximately a year and send close-up photographs of the lunar surface during that time. It will also conduct mapping by radar and other instruments to study the lunar atmosphere and fluctuation of the sun's radiation.

This failure denotes the second crash landing on the Moon for India's Chandrayaan landers. Undeterred, India has already announced a third lander mission in 2024 as it strives to become only the fourth nation to land on the Moon. Never the less, the failure is a huge setback to the Indian space program.

NASA stated that out of 109 attempted lunar missions by space agencies, 48 have failed.

This picture is a screen grab taken from a live webcast by the Indian Space Research Organization (ISRO) and shows the Vikram lander descending to its eventual crash-landing on the surface of the Moon. (Photo from AFP News Agency).

Canada's debut into space came with the launch of its first satellite Alouette 1 in 1962. It currently operates a fleet of RADARSAT and SCISAT earth observation satellites, ANIK communications satellites, and BRITE science satellites as well as micro and hybrid spacecraft.

The Canadian space program is controlled by the Canadian Space Agency (CSA). The nation currently possesses no indigenous launch system, and depends on the US, India and Russia to launch its spacecraft.

The successful launch of the Azur satellite in 1969 demonstrated Germany's space-faring capabilities to the world. Germany has launched several spacecraft including telecommunications, navigation and earth observation satellites, and is involved in core missions such as Cassini-Huygens mission to Saturn and its moons, European space laboratory Columbus, Dawn mission to Vesta and Ceres and the Galileo navigation system.

The national space program is implemented by the German Aerospace Center (DLR) which supports the German space industry to meet the strategic goals in the European programs associated with the ESA and the European Organization for the Exploitation of Meteorological Satellites.

The French space program constitutes of both civil and military space missions and its space policy is implemented by state-owned Centre National Space Studies (CNES), which is responsible for the development and execution of space programs alongside industry and the scientific community. The nation's in-orbit spacecraft constitute earth observation and reconnaissance satellites, electronic signals intelligence satellites, civil and military communications satellites.

France is one of the largest contributors to the European Space Agency (ESA) which is headquartered in Paris. The space research and development is carried out at Toulouse Space Centre whereas CNES, ESA and Arianespace conduct launches from the Guiana Space Centre.

Luxembourg operates a large number of communications and remote sensing satellites, making it one of the top nations with space existence. It is one of the Member States of ESA and carries space research activities under its National Action Plan for Space R&D. The nation is also home to the headquarters of world's leading telecommunications satellite operators SES European Society of Satellites and Intelsat.

The Luxembourg Space Cluster unites highly specialized companies and government research agencies focusing on space telecommunications, global navigation satellite system and location-based services, earth observation, maritime safety and protection, and space technologies.

The Leading side of the Moon (NASA)

The next American attempt to put a lander on the Moon is scheduled in 2021. An Astrobotic Technology robot spacecraft will launch atop a United Launch Alliance Vulcan rocket. It will be the first of twenty missions NASA has scheduled to explore the lunar surface over the next decade.

If successful the mission will mark the first U.S. landing on the Moon since the Apollo program.

The company, out of Pittsburgh, was selected by the U.S. space agency's Commercial Lunar Payload Services (CLPS) program to deliver 14 payloads to the Moon with its Peregrine lunar lander. This was a second awarded contract and Astrobotic now has 28 payload deliveries under contract for future missions.

Astrobotic's Peregrine moon lander, (Photo courtesy NASA Goddard Spaceflight Center).

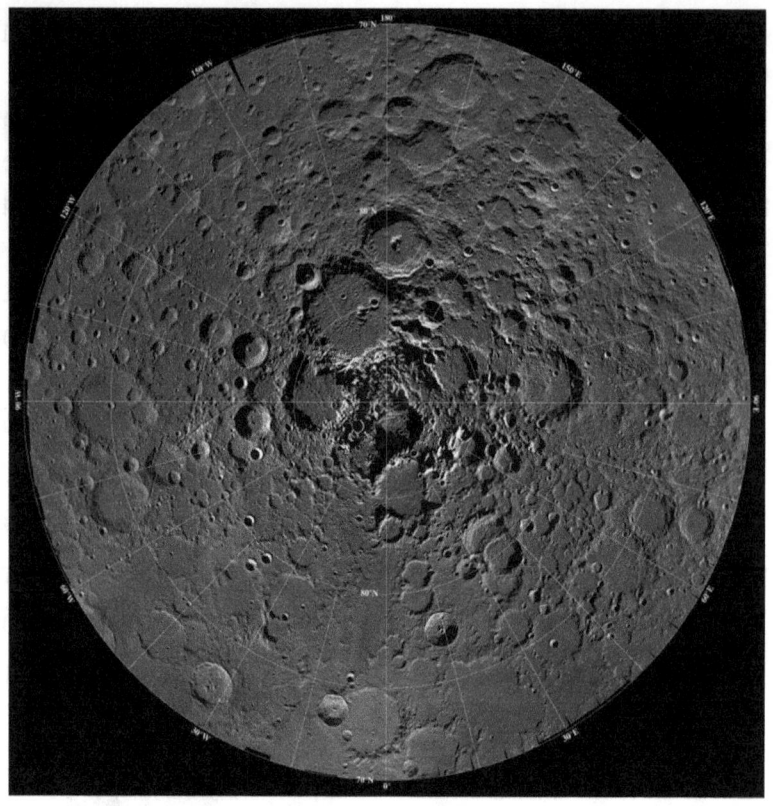

The north pole of the Moon. (NASA)

There will be immediate challenges and counter claims made from space faring nations and that is without a doubt.

The reality is we may not overcome all the problems inherent in going to or colonizing Mars in the foreseeable future. However, I am confident we can do so **on the Moon**. Indeed, if we miss our chance, the world may condemn us for our timidity should the Moon fall into the wrong hands.

Never again will we have a better claim to the Moon. Let us accept the challenge and embrace what is obviously our, and all of humanity's, destiny.

Chapter 11
Threats to U.S. Space Programs

Air Force Issues China, Russia Warning

In early December of 2019 the U.S. Air Force's chief of staff, General David Goldfein, warned that the greatest threat to the U.S. in space was Russia and Communist China.

"Russia is a rather dangerous threat because it's an economy in decline and the demographics [of that] are challenging for Vladimir Putin. But China is the face of the threat. China has the economy [to threaten us]."

Both of these adversarial countries have developed electronic jamming, rockets and lasers that can destroy or blind our satellites. They are also working on maneuverable satellite types of spacecraft capable of tampering with American spacecraft.

"We have to defend what we have… At the same time we need to transition to defendable architecture. But it's not good enough to take punches in the ring. At some point you have to punch back. And third, we need to transition our force into a war fighting force, as opposed to one that's accustomed to operating spacecraft in a 'relatively benign' environment."

The general went on to add "In every war game, we determined that if you move first in space, you're not guaranteed to win. But if you move second, you're likely to lose."

Communist Chinese Hacking and Espionage

For decades the Chinese Communists have been stealing American technology through any means including trade, innocent appearing eavesdropping products, subversion of people, direct spying, and the hacking of our computing systems and spacecraft. They are even shriveling our children through attractive gaming devices.

One of the latest examples of this is the grounding of all U.S. government owned drones made in China, and even drones repaired with parts from China because of cybersecurity concerns. Most glaring at this time is the suppression of the Chinese communications company of Huawei because the system's infrastructure equipment may allow surveillance of users by the Chinese government.

Huawei is the center of espionage allegation over the Chinese 5G network equipment that is seen as a Trojan horse. The U.S. passed a defense funding bill that barred the federal government from doing business with Huawei as well as ZTE, and Chinese surveillance products because of security questions.

The International Space Station (ISS). NASA photo.

Chinese Hackers Take Control of ISS

In 2012 the International Space Station was hacked by China and the maneuvering rockets were fired by them for over a minute, moving the ISS into a dangerous orbit. This orbit put the station in jeopardy of possibly being hit by space debris. It was a warning, a warning we apparently did not take seriously. That's obvious in the former vice president's statement I alluded to earlier.

One reason for the many hacks is lost laptop computers by NASA employees or contractors with unencrypted codes easily used for hacking many NASA projects. Another reason is the stealing of employee's identification badges and entry keys.

The Communist Chinese are not good folks, they are not honorable in their dealings with us, they lie, they cheat, and they steal. All to one end, they intend to replace us; they intend to make us irrelevant in all things and especially in the exploration of space.

The ISS hacking was not the only cyber strike they carried out. They also gained entry to the Jet Propulsion Laboratory, one of the most important of NASA cyber sites.

Three years later a Chinese-born American scientist working for NASA on the Mars rover project, Dr. Rongxing Li suddenly resigned and promptly disappeared. He possessed a 'secret' clearance and worked on several sensitive projects including U.S. military missiles.

Stating that his parents were ill the Chinese scientist flew to China. He was soon followed by his wife who was found to be carrying USB devices containing U.S. sensitive data. Amazingly neither she nor Rongxing Li were ever charged with espionage. What genius government administrator made that decision?

If anyone was listening in America they could probably hear the laughing in China. They played us for fools, and recently, they did it again.

Chinese-born Wei Sun, employed by Raytheon Missile Systems as an engineer in 2019 informed the company he was going overseas on a trip. He also told them he was taking his work related laptop with him.

Like Rongxing Li, he held a secret clearance and worked on sensitive missile projects. Instead of taking custody of his work laptop Raytheon simply warned him not to take it with him.

But of course, he did. While still overseas he connected to Raytheon's internal network using the laptop and most assuredly downloaded sensitive information. Then he resigned from the job he had occupied for ten years. A week later Sun foolishly returned to the U.S. and was confronted by Raytheon security officials (finally) and after confirming that sensitive data was still on the laptop he was arrested by the FBI.

The court case is continuing. This is only the tip of the ice berg on Communist Chinese espionage in the U.S. Can there be any doubt about what China is trying to do? No Joe Biden, these people are not our friends!

Scientist Hides China Links from NASA

Professor Anming Hu of the University of Tennessee was recently discovered (February of 2020) to be hiding his links to Beijing University of Technology from Tennessee Knoxville University officials directly and from NASA indirectly. According to court documents NASA would not have funded Hu's work had they known of the affiliation with the Chinese university.

The Scheme to Defraud NASA

21. Beginning in 2016 HU engaged in a scheme to defraud NASA by falsely representing and concealing his affiliation with BJUT to UKT. Through his fraudulent representations and omissions to UKT about his affiliation with BJUT,

HU knowingly and willfully caused UTK to falsely certify to NASA and to NASA contractors that UTK was in compliance with NASA's China Funding Restriction regarding NASA-funded projects that UKT sought and obtained on HU's behalf. Had Hu fully disclosed to UTK, his affiliation with BJUT, UTK would not have certified to NASA and NASA contractors that UTK was in compliance with NASA's China Funding Restriction and NASA would not have awarded NASA-funded projects to HU.

Hu has been detained pending a court appearance in the first week of March 2020. Hu was working a Mars sample project for NASA.

This recent incident makes it apparent that Communist China is continuing to conduct espionage against the United States today.

Russians Threaten U.S. Satellite

"We view this behavior as unusual and disturbing,"

–General John Raymond, U.S. Space Force

Early in February a sharp eyed student and amateur satellite tracker at Purdue University noted the fact that a Russian Cosmos 2542 *Inspector* satellite had made a tactical move. It closed the distance between it and a National Reconnaissance Office (NRO) satellite dubbed USA 245 to less than two hundred miles. It then began shadowing the 245, likely a ground surveillance satellite. This kind of maneuver could be construed as an attempt at espionage, or a prelude to attack, although the former is more likely at this time. But, there will come a time when such a move will likely be an attack or the preparation for one.

USA 245 is an older satellite relative to Cosmos 2542. It has been in orbit for six years whereas the Russian craft is only about three months old. The *Inspector* adjusted its variable position to approach USA 245 closer during daylight periods and further away during the hours of darkness. It may be analyzing the American satellite's camera position determine what it's looking at back on Earth.

On February 10 the trailing satellite launched a second spacecraft that joined alongside the first and in an ominous maneuver; both are continuing to shadow the U.S. satellite.

This initial report on a foreign satellite intercepting a U.S. satellite is the first of its kind and is in keeping with the stated mission of the Space Force.

The Space Force is ear tagged for a $15.4 billion budget increase in this year's federal government funding allocation. These funds will help the new military branch get the technology and personnel needed to carry out their mission of protecting U.S. interests both physically and in accordance with stated U.S. policies, in outer space.

Of course the two repressive government's space programs were an important factor in the founding of the USSF. Part of the Space Force's mission is to neutralize interference and defeat aggressive acts in defense of our own efforts and activities in space.

The ability to spy on other satellites is not unique to Russia but is also done by the U.S. and the Chinese.

This may not look like *Star Wars* yet, but it may be its precursor.

Prediction; a China Surprise Moon Landing

Recently a space aficionado asked "What the hell are the Chinese doing on the far side of the Moon?" I think I know and I'm willing to make a prediction about that.

Recently Communist China has been beset by problems that have somewhat reduced their economy including the tariff war with the U.S. and the outbreak of the corona virus. These and other events may serve to slow down the funding of their space program. But I wouldn't bet on it.

The recent launch of their Long March 5 rocket and the success of their Chang'e lunar rover currently on the dark side of the Moon show they remain quite active in their space efforts.

They are well aware of our intent to return to the Moon in 2024 and I believe they have more goals than just landing a manned mission on the Moon. I think they will launch a surprise and maybe even a secret manned mission before we launch ours.

Their reasons to do this are threefold. The first is to show their space program is superior to ours by beating us to the Moon in this new, 21st century space race. Their other two reasons complement one another. They include landing on the lunar South Pole as near as they can to the water source there, and set up a permanent base. If they succeed in that they will be the very first nation to do so. And then, they intend to stake a claim, likely not to the entire Moon, but surely for the area surrounding and including the frozen water source.

If they do this, they will muddy the waters on any claim we subsequently make on the Moon. They will be firmly ensconced and will no doubt become a major wrench thrown into our own plans for the Moon. This is why we cannot wait, but must make our claim as soon as possible and follow it as soon as possible with a manned mission to the Moon, and the training of our Space Force.

Too many times in our nation's past have we been surprised by our adversaries. Let's be prepared for this challenge to our supremacy in space!

When will the Chinese launch this surprise Moon mission? I don't know, but suffice it to say they will do it as soon as they feel they have a good chance of success. At the latest however, I think they will try to steal a march on us at least a year before we are ready to respond. In that case, they could have an entire year on the Moon before we arrive and thus bolster their own claim.

Remember, they are only loosely subject to the opinions of their people. They have no problem gambling the lives of a few astronauts to gain a real and lasting foothold on the Moon. They are willing to do this because they understand all too well that the Moon is the key to the solar system's riches.

Chapter 12
The Moon, Gateway to the Solar System and Beyond

The Moon is our gateway to the solar system and beyond. It is on the Moon we can ever really hope to construct spaceships large enough to safely travel to Mars and the other places we want to go to and stay.

On the lunar surface and in orbit around the Moon our astronauts both civilian and Space Force can learn. They can learn about space, how to move in space both in spacecraft and on a low gravity surface. Valuable lessons in the safe operation of not only spacecraft but human carrying rovers, operation of space locks to enter and exit habitats and equipment. We can take our time to learn about power production, water extraction from various sources, and hydroponic food production in low gravity. And many, many other vital and necessary tasks needed to survive in space, on the lunar and Mars surface, and under the surface in volcanic tunnels. This will well and truly prepare us for doing the same things on Mars, asteroids, and even the moons of Jupiter and Saturn.

I'm concerned that if we try to take a Mars Direct approach we could be courting a space disaster that could set us back for years. I especially have reservations about going for broke to send colonists intended upon remaining on our first trip to Mars.

The possibility exists that in the end, the Moon may be the only other space body that we humans ever occupy in high numbers and where we can establish a second world civilization.

We are making great strides in spacecraft design and production, in habitat construction, fuel and oxygen production as well as water on the Moon. Let's put that knowledge to work on a place we can easily reach in a matter of days, eventually a matter of hours in case of a space emergency.

We cannot say the same if we venture too far away too soon. No, the Moon is our key and our stepping stone to the greater solar system. Let us do this first.

NASA illustration of a printer produced lunar habitat that could be made from available materials on the surface of the Moon.

Military spacecraft training as well as weapons development could take place on the farside of the Moon out of the view of prying eyes. In fact I think that initially the farside should be kept almost exclusively for military use. Yes, I'm recommending a new Area 51 on the Moon. We need an area designated for the Space Force to hone its skills, develop tactics, and allow for realistic training.

Without a doubt our adversaries will conduct surveillance from a distance of our farside activities. This will give us a chance to experiment and learn space camouflage both for operating spacecraft in space and hiding on and below the lunar surface.

The thicker crust on the Moon's farside, up to thirty miles deeper than the nearside, would suggest it's a better place for space weapons testing.

NASA illustration of a Moon orbiting space habitat similar to the ISS.

Chapter 13
The World's Reaction

Our primary space faring opponents, Communist China and the Russian Federation will, without a doubt, file motions in the United Nations to rescind our claim. However, the UN lacks a space force or indeed any military or police power to stop us.

If we have already reached out to assure our space allies that they will be allowed to continue their lunar activities, albeit under U.S. supervision, they will likely not seek redress through the UN.

You may recall I mentioned that both our two rivals also have a space force. Almost assuredly they will try to quickly launch manned lunar missions with the intent of planting their flag on the surface of the Moon and laying their own claim.

I think the rest of the world will feel it's right and proper for the U.S. to dominate on the Moon as we have for the past half century and more.

You Will Remember Me for Centuries

Some Legends are told
Some turn to dust or to gold
But you will remember me
Remember me for centuries

These are some of the lyrics of the very popular song by the band 'Fall Out Boy' *Centuries*. Whoever the national leader is that announces the American claim to the Moon and succeeds in making it stick, will become the person remembered for centuries. Indeed, that person will likely be remembered as long as humans exist.

So, who are the candidates for such a thing, who has the necessary hutzpah to pull this off? At the head of the list has to be President Donald Trump. And, if he chose to do this I think he would not only insure his legendary immortality, but insure his reelection this fall, assuming he does it right now. And, he has the power and stroke behind him to pull it off. But are there other candidates?

There are some visionary space leaders out there, individuals of great wealth and stature that might be bold enough to state they will claim the Moon on behalf of the United States. And I think they would not only have the right, especially if it's their spaceship that takes our astronauts back to the Moon, to do just that.

Individuals that come to mind from the commercial sector are Elon Musk and Jeff Bezos. From NASA comes their national director Jim Bridenstine, my fellow Oklahoman. Also with a background in NASA is outspoken Apollo astronaut Edwin 'Buzz' Aldrin, the most senior astronaut still living to have walked on the Moon.

There are other Apollo astronauts that have visited the Moon's surface besides Buzz who remain. They include David Scott, Charles Duke, and Harrison Schmitt. These men, as well as their partners overhead in the Apollo command module are obviously some of our greatest space heroes.

Those who orbited the Moon in the command module include Frank Borman, Jim Lovell, Bill Anders, Tom Stafford, Michael Collins, Fred Haise, Ken Mattingly, and Al Worden.

All these men are of advanced age, yet their opinions still carry a good deal of weight in space circles. Were they to sign a petition advocating a claim to the Moon, I have little doubt their wishes would be honored by President Trump.

We currently have 38 active members of the astronaut corps, I have little doubt but that a majority of them would agree to America laying claim to the Moon. They could join with the Apollo astronauts and any others I have failed to mention in signing the petition if they are brave enough.

There are thousands of space workers and rocket scientists across the nation that I believe would support our claim to the Moon.

General John Raymond, the commander of the U.S. Space Force could easily recommend and justify that it is incumbent upon the United States to lay claim to the Moon. While he is doing that he can outline what he thinks the Space Force needs to immediately safeguard and defend America's newest territory, the Moon.

There are others, captains of industry who can envision a bright and prosperous future for a civilization to grow upon the Moon. A civilization founded by the free people of the United States of America. They can work toward establishing our claim through their moral and financial support of such a move.

Let's do this thing for all of America, for all of humanity. Let us boldly go forward and embrace our destiny in outer space!

This is not the end; in fact it is only the beginning of my intense look at us claiming the Moon. You can look forward to my coming book *Claim the Moon II*. The working cover is below.

Epilogue

CLAIM THE MOON FOR THE U.S.A. -NATIONAL MOVEMENT-

SEND TO: Our president, our senators, our representatives and our state governors and state legislatures.

The time has come for us to claim what belongs to the people of the United States of America, earned by the hard work of our government agencies, the dedication of our space workers, and especially the sacrifices of our courageous astronauts who remain the only humans to have <u>ever landed and walked upon</u> *THE MOON*!

We must take advantage of our technological edge in the exploration of space and the Moon. We must stop sharing our hard won scientific knowledge with the repressive governments of Russia and the Communist Chinese. We must remain foremost among the space faring nations and increase our presence exponentially on the Moon! To do this we must lay our claim to the Moon, the entire Moon, and we must do it right now!

We must be prepared to defend that claim by force if necessary by employment of the U.S. Space Force. If you need information on how to do this I recommend you read the book *Claim the Moon* which outlines how to checkmate China, recover every cent we have spent on space exploration, and guarantee our future prosperity! I am a citizen of the U.S.A.

Online: www.lulu.com/greenpheon7

_____
_____
Signature, printed name, address.

NATIONAL MOVEMENT LETTER TO CLAIM THE MOON

Copy, sign it, print your name and address and send to the leader(s) or media of your choice. Add your personal comments on the back. Do something good for your nation today! Future generations will thank you and I thank you.

Remembering the Challenger Crew Then and Now

Even as we look forward to resuming manned spaceflight and returning to the Moon, we are reminded of our past.

On January 28, 1986 I was stationed at Fort Bliss, Texas for the purpose of attending the U.S. Army Sergeants Major Academy (USASMA).

There, I studied international affairs, world conflict, the Cold War, and military subjects such as the role of a command sergeant major.

At the time of my attendance, only the top four percent of senior non-commissioned officers of the Army, the Navy and Marines, the U.S. Air Force, and the U.S. Coast Guard were considered for attendance as a resident at the academy.

Like other educational institutions, the academy published a class book for each academy class. We were class XXVII of USASMA to graduate and we dedicated our class book to the astronauts that perished in the Challenger.

Shown in the picture below, left to right starting in the back row are; Mission Specialist Ellison Onizuka, Teacher in Space Christa McAuliffe, Payload Specialist Greg Jarvis, Mission Specialist Judy Resnik, Space Shuttle Pilot Mike Smith, Commander Dick Scobee, and Mission Specialist Ron McNair.

On an unusually cold morning in Florida, the seven crew members of Space Shuttle Challenger were killed when their spaceship suddenly exploded less than two minutes after launch.

Although it was the tenth launch of Challenger, the cold morning caused the 'O' rings of the rocket engines to leak, resulting in the explosion and breakup of the shuttle.

IN MEMORY OF . . .

Shortly after we began our classes at the academy, we were shocked by a tragedy that none of us thought possible. On January 28, Shuttle mission 51-L (Space Shuttle Challenger) lifted off from launch pad 39B of the Kennedy Space Center at precisely 11:30 AM. The Challenger carried a crew of seven astronauts and the Tracking and Data Relay Satellite. 73 seconds after launch, the solid rocket booster that was carrying the shuttle spaceward exploded, killing all aboard. Many of us in Class 27 grew up with the space program, and like many Americans, took for granted the courage of the men and women who have served their nation in the vacuum of space. We don't any longer.

This book is dedicated to the memory of the astronauts of Challenger, pictured below. (Back row, left to right) Mission Specialist El Onizuka, Teacher in Space Participant S. Christa McAuliffe, Payload Specialist Greg Jarvis, and Mission Specialist Judy Resnik; (front row, left to right) Pilot Mike Smith, Commander Dick Scobee, and Mission Specialist Ron McNair. (Photos courtesy of the National Aeronautics and Space Administration)

The Space Shuttle program lasted from 1981 to 2011 and no American has launched in a U.S. Spacecraft since that time. That should change this year.

Selected remains of the Challenger are memorialized at Cape Canaveral. NASA photo.

About the Author

Walt Cross is retired from the U.S. Army. He is a *Pheon* veteran of the Vietnam War and served with Battery A, 1st Battalion, 7th U.S. Artillery, 1st Infantry Division from 1968 to 1970.

Including *Claim the Moon*, Walt has authored fifteen books on history and science and two novels.

Walt holds university degrees in both history and science. Additionally, he is a graduate of the U.S. Army Sergeants Major Academy (Fort Bliss, Texas). He is also the Executive Director emeritus of the Oklahoma ESGR (Employer Support of the Guard and Reserve) a Department of Defense soldier support program.

In addition to writing and editing books, Walt is an illustrator. He designs the covers of all of his books to include *Claim the Moon*.